tredition

tredition was established in 2006 by Sandra Latusseck and Soenke Schulz. Based in Hamburg, Germany, tredition offers publishing solutions to authors and publishing houses, combined with worldwide distribution of printed and digital book content. tredition is uniquely positioned to enable authors and publishing houses to create books on their own terms and without conventional manufacturing risks.

For more information please visit: www.tredition.com

TREDITION CLASSICS

This book is part of the TREDITION CLASSICS series. The creators of this series are united by passion for literature and driven by the intention of making all public domain books available in printed format again - worldwide. Most TREDITION CLASSICS titles have been out of print and off the bookstore shelves for decades. At tredition we believe that a great book never goes out of style and that its value is eternal. Several mostly non-profit literature projects provide content to tredition. To support their good work, tredition donates a portion of the proceeds from each sold copy. As a reader of a TREDITION CLASSICS book, you support our mission to save many of the amazing works of world literature from oblivion. See all available books at www.tredition.com.

 Project Gutenberg

The content for this book has been graciously provided by Project Gutenberg. Project Gutenberg is a non-profit organization founded by Michael Hart in 1971 at the University of Illinois. The mission of Project Gutenberg is simple: To encourage the creation and distribution of eBooks. Project Gutenberg is the first and largest collection of public domain eBooks.

Time and Tide A Romance of the Moon

Robert S. (Robert Stawell), Sir Ball

Imprint

This book is part of TREDITION CLASSICS

Author: Robert S. (Robert Stawell), Sir Ball
Cover design: Buchgut, Berlin – Germany

Publisher: tredition GmbH, Hamburg - Germany
ISBN: 978-3-8472-3915-4

www.tredition.com
www.tredition.de

Copyright:
The content of this book is sourced from the public domain.

The intention of the TREDITION CLASSICS series is to make world literature in the public domain available in printed format. Literary enthusiasts and organizations, such as Project Gutenberg, worldwide have scanned and digitally edited the original texts. tredition has subsequently formatted and redesigned the content into a modern reading layout. Therefore, we cannot guarantee the exact reproduction of the original format of a particular historic edition. Please also note that no modifications have been made to the spelling, therefore it may differ from the orthography used today.

THE ROMANCE OF SCIENCE.

TIME AND TIDE,
A Romance of the Moon.

BEING TWO LECTURES DELIVERED
IN THE THEATRE OF THE LONDON INSTITUTION,
ON THE AFTERNOONS OF NOVEMBER 19 AND 26, 1888.

BY

SIR ROBERT S. BALL, LL.D., F.R.S.,

ROYAL ASTRONOMER OF IRELAND.
PUBLISHED UNDER THE DIRECTION OF THE COMMITTEE OF
GENERAL LITERATURE AND EDUCATION APPOINTED
BY THE SOCIETY FOR PROMOTING CHRISTIAN
KNOWLEDGE.

View of the Moon two days after first quarter.
From a photograph by Mr. Lewis M. Rutherford.
Frontispiece.

TO
The Members of the London Institution
I DEDICATE
THIS LITTLE BOOK.

PREFACE.

Having been honoured once again with a request that I should lecture before the London Institution, I chose for my subject the Theory of Tidal Evolution. The kind reception which these lectures received has led to their publication in the present volume. I have taken the opportunity to supplement the lectures as actually delivered by the insertion of some additional matter. I am indebted to my friends Mr. Close and Mr. Rambaut for their kindness in reading the proofs.

Robert S. Ball.

Observatory, Co. Dublin,
April 26, 1889.

TIME AND TIDE. [9]

LECTURE I.

It is my privilege to address you this afternoon on a subject in which science and poetry are blended in a happy conjunction. If there be a peculiar fascination about the earlier chapters of any branch of history, how great must be the interest which attaches to that most primeval of all terrestrial histories which relates to the actual beginnings of this globe on which we stand.

In our efforts to grope into the dim recesses of this awful past, we want the aid of some steadfast light which shall illumine the dark places without the treachery of the will-o'-the-wisp. In the absence of that steadfast light, vague conjectures as to the beginning of things could never be [10] entitled to any more respect than was due to mere matters of speculation.

Of late, however, the required light has been to some considerable extent forthcoming, and the attempt has been made, with no little success, to elucidate a most interesting and wonderful chapter of an exceedingly remote history. To chronicle this history is the object of the present lectures before this Institution.

First, let us be fully aware of the extraordinary remoteness of that period of which our history treats. To attempt to define that period chronologically would be utterly futile. When we have stated that it is more ancient than almost any other period which we can discuss, we have expressed all that we are really entitled to say. Yet this conveys not a little. It directs us to look back through all the ages of modern human history, through the great days of ancient Greece and Rome, back through the times when Egypt and Assyria were names of renown, through the days when Nineveh and Babylon were mighty and populous cities in the zenith of their glory. Back earlier still to those more ancient nations of which [11] we know hardly anything, and still earlier to the prehistoric man, of whom we know less; back, finally, to the days when man first trod on this planet, untold ages ago. Here is indeed a portentous retrospect from most points of view, but it is only the commencement of that which our subject suggests.

For man is but the final product of the long anterior ages during which the development of life seems to have undergone an exceedingly gradual elevation. Our retrospect now takes its way along the vistas opened up by the geologists. We look through the protracted tertiary ages, when mighty animals, now generally extinct, roamed over the continents. Back still earlier through those wondrous secondary periods, where swamps or oceans often covered what is now dry land, and where mighty reptiles of uncouth forms stalked and crawled and swam through the old world and the new. Back still earlier through those vitally significant ages when the sunbeams were being garnered and laid aside for man's use in the great forests, which were afterwards preserved by being transformed into seams of coal. Back still earlier [12] through endless thousands of years, when lustrous fishes abounded in the oceans; back again to those periods characterized by the lower types of life; and still earlier to that incredibly remote epoch when life itself began to dawn on our awakening globe. Even here the epoch of our present history can hardly be said to have been reached. We have to look through a long succession of ages still antecedent. The geologist, who has hitherto guided our view, cannot render us much further assistance; but the physicist is at hand—he teaches us that the warm globe on which life is beginning has passed in its previous stages through every phase of warmth, of fervour, of glowing heat, of incandescence, and of actual fusion; and thus at last our retrospect reaches to that particular period of our earth's past history which is specially illustrated by the modern doctrine of Time and Tide.

The present is the clue to the past. It is the steady application of this principle which has led to such epoch-making labours as those by which Lyell disclosed the origin of the earth's crust, Darwin the origin of species, Max Müller the origin [13] of language. In our present subject the course is equally clear. Study exactly what is going on at present, and then have the courage to apply consistently and rigorously what we have learned from the present to the interpretation of the past.

Thus we begin with the ripple of the tide on the sea-beach which we see to-day. The ebb and the flow of the tide are the present manifestations of an agent which has been constantly at work. Let that present teach us what tides must have done in the indefinite past.

It has been known from the very earliest times that the moon and the tides were connected together—connected, I say, for a great advance had to be made in human knowledge before it would have been possible to understand the true relation between the tides and the moon. Indeed, that relation is so far from being of an obvious character, that I think I have read of a race who felt some doubt as to whether the moon was the cause of the tides, or the tides the cause of the moon. I should, however, say that the moon is not the sole agent engaged in producing this [14] periodic movement of our waters. The sun also arouses a tide, but the solar tide is so small in comparison with that produced by the moon, that for our present purpose we may leave it out of consideration. We must, however, refer to the solar tide at a later period of our discourses, for it will be found to have played a very splendid part at the initial stage of the Earth-Moon History, while in the remote future it will again rise into prominence.

It will be well to set forth a few preliminary figures which shall explain how it comes to pass that the efficiency of the sun as a tide-producing agent is so greatly inferior to that of the moon. Indeed, considering that the sun has a mass so stupendous, that it controls the entire planetary system, how is it that a body so insignificant as the moon can raise a bigger tide on the ocean than can the sun, of which the mass is 26,000,000 times as great as that of our satellite?

This apparent paradox will disappear when we enunciate the law according to which the efficiency of a tide-producing agent is to be estimated. This law is somewhat different from the familiar form in [15] which the law of gravitation is expressed. The gravitation between two distant masses is to be measured by multiplying these masses together, and dividing the product by the *square* of the distance. The law for expressing the efficiency of a tide-producing agent varies not according to the inverse square, but according to the inverse *cube* of the distance. This difference in the expression of the law will suffice to account for the superiority of the moon as a tide-producer over the sun. The moon's distance on an average is about one 386th part of that of the sun, and thus it is easy to show that so far as the mere attraction of gravitation is concerned, the efficiency of the sun's force on the earth is about one hundred and seventy-five times as great as the force with which the moon attracts the earth. That is of course calculated under the law of the inverse square. To determine the tidal efficiency we have to divide this by three hundred and eighty-six, and thus we see that the tidal efficiency of the sun is less than half that of the moon.

When the solar tide and the lunar tide are acting in unison, they conspire to produce very high [16] tides and very low tides, or, as we call them, spring tides. On the other hand, when the sun is so placed as to give us a low tide while the moon is producing a high tide, the net result that we actually experience is merely the excess of the lunar tide over the solar tide; these are what we call neap tides. In fact, by very careful and long-continued observations of the rise and fall of the tides at a particular port, it becomes possible to determine with accuracy the relative ranges of spring tides and neap tides; and as the spring tides are produced by moon plus sun, while the neap tides are produced by moon minus sun, we obtain a means of actually weighing the relative masses of the sun and moon. This is one of the remarkable facts which can be deduced from a prolonged study of the tides.

The demonstration of the law of the tide-producing force is of a mathematical character, and I do not intend in these lectures to enter into mathematical calculations. There is, however, a simple line of reasoning which, though it falls far short of actual demonstration, may yet suffice to give a plausible reason for the law. [17]

The tides really owe their origin to the fact that the tide-producing agent operates more powerfully on those parts of the tide-exhibiting body which are near to it, than on the more distant portions of the same. The nearer the two bodies are together, the larger proportionally will be the differences in the distances of its various parts from the tide-producing body; and on this account the leverage, so to speak, of the action by which the tides are produced is increased. For instance, if the two bodies were brought within half their original distance of each other, the relative size of each body, as viewed from the other, will be doubled; and what we have called the leverage of the tide-producing ability will be increased twofold. The gravitation also between the two bodies is increased fourfold when the distance is halved, and consequently, the tide-producing ability is doubled for one reason, and increased fourfold again by another; hence, the tides will be increased eightfold when the distance is reduced to one half. Now, as eight is the cube of two, this illustration may be taken as a verification of the law, that the efficiency of a body as a tide-producer varies [18] inversely as the cube of the distance between it and the body on which the tides are being raised.

For simplicity we may make the assumption that the whole of the earth is buried beneath the ocean, and that the moon is placed in the plane of the equator. We may also entirely neglect for the present the tides produced by the sun, and we shall also make the further assumption that friction is absent. What friction is capable of doing we shall, however, refer to later on. The moon will act on the ocean and deform it, so that there will be high tide along one meridian, and high tide also on the opposite meridian. This is indeed one of the paradoxes by which students are frequently puzzled when they begin to learn about the tides. That the moon should pull the water up in a heap on one side seems plausible enough. High tide will of course be there; and the student might naturally think that the water being drawn in this way into a heap on one side, there will of course be low tide on the opposite side of the earth. A natural assumption, perhaps, but nevertheless a very wrong one. There are at every moment two opposite parts of the earth in [19] a condition of high water; in fact, this will be obvious if we remember that every day, or, to speak a little more accurately, in every twenty-four hours and fifty one minutes, we have on the average two high tides at each locality. Of course this could not be if the moon raised only one heap of high water, because, as the moon only appears to revolve around the earth once a day, or, more accurately, once in that same average period of twenty-four hours and fifty-one minutes, it would be impossible for us to have high tides succeeding each other as they do in periods a little longer than twelve hours, if only one heap were carried round the earth.

The first question then is, as to how these two opposite heaps of water are placed in respect to the position of the moon. The most obvious explanation would seem to be, that the moon should pull the waters up into a heap directly underneath it, and that therefore there should be high water underneath the moon. As to the other side, the presence of a high tide there was, on this theory, to be accounted for by the fact that the moon pulled the earth away from the waters on the more [20] remote side, just as it pulled the waters away from the more remote earth on the side underneath the moon. It is, however, certainly not the case that the high tide is situated in the simple position that this law would indicate, and which we have represented in Fig. 1, where the circular body is the earth, the ocean surrounding which is distorted by the action of the tides.

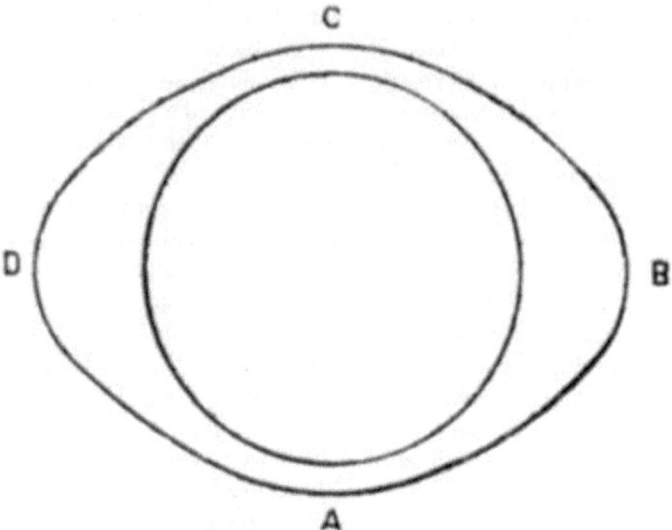

Fig. 1

We have here taken an oval to represent the shape into which the water is supposed to be forced or drawn by the tidal action of the tide-producing body. This may possibly be a correct representation of what would occur on an ideal [21] globe entirely covered with a frictionless ocean. But as our earth is not covered entirely by water, and as the ocean is very far from being frictionless, the ideal tide is not the tide that we actually know; nor is the ideal tide represented by this oval even an approximation to the actual tides to which our oceans are subject. Indeed, the oval does not represent the facts at all, and of this it is only necessary to adduce a single fact in demonstration. I take the fundamental issue so often debated, as to whether in the ocean vibrating with ideal tides the high water or the low water should be under the moon. Or to put the matter otherwise; when we represent the displaced water by an oval, is the long axis of the oval to be turned to the moon, as generally supposed, or is it to be directed at right angles therefrom? If the ideal tides were in any degree representative of the actual tides, so fundamental a question as this could be at once answered by an appeal to the facts of observation. Even if friction in some degree masked the phenomena, surely one would think that the state of the actual tides should still enable us to answer this question. [22]

But a study of the tides at different ports fails to realize this expectation. At some ports, no doubt, the tide is high when the moon is on the meridian. In that case, of course, the high water is under the moon, as apparently ought to be the case invariably, on a superficial view. But, on the other hand, there are ports where there is often low water when the moon is crossing the meridian. Yet other ports might be cited in which every intermediate phase could be observed. If the theory of the tides was to be the simple one so often described, then at every port noon should be the hour of high water on the day of the new moon or of the full moon, because then both tide-exciting bodies are on the meridian at the same time. Even if the friction retarded the great tidal wave uniformly, the high tide on the days of full or change should always occur at fixed hours; but, unfortunately, there is no such delightful theory of the tides as this would imply. At Greenock no doubt there is high water at or about noon on the day of full or change; and if it could be similarly said that on the day of full or change there was high water everywhere at local noon, then [23] the equilibrium theory of the tides, as it is called, would be beautifully simple. But this is not the case. Even around our own coasts the discrepancies are such as to utterly discredit the theory as offering any practical guide. At Aberdeen the high tide does not appear till an hour later than the doctrine would suggest. It is two hours late at London, three at Tynemouth, four at Tralee, five at Sligo, and six at Hull. This last port would be indeed the haven of refuge for those who believe that the low tide ought to be under the moon. At Hull this is no doubt the case; and if at all other places the water behaved as it does at Hull, why then, of course, it would follow that the law of low water under the moon was generally true. But then this would not tally with the condition of affairs at the other places I have named; and to complete the cycle I shall add a few more. At Bristol the high water does not get up until seven hours after the moon has passed the meridian, at Arklow the delay is eight hours, at Yarmouth it is nine, at the Needles it is ten hours, while lastly, the moon has nearly got back to the meridian again ere it has succeeded in dragging [24] up the tide on which Liverpool's great commerce so largely depends.

Nor does the result of studying the tides along other coasts beside our own decide more conclusively on the mooted point. Even ports in the vast ocean give a very uncertain response. Kerguelen Island and Santa Cruz might seem to prove that the high tide occurs under the moon, but unfortunately both Fiji and Ascension seem to present us with an equally satisfactory demonstration, that beneath the moon is the invariable home of low water.

I do not mean to say that the study of the tides is in other respects such a confused subject as the facts I have stated would seem to indicate. It becomes rather puzzling, no doubt, when we compare the tides at one port with the tides elsewhere. The law and order are, then, by no means conspicuous, they are often hardly discernible. But when we confine our attention to the tides at a single port, the problem becomes at once a very intelligible one. Indeed, the investigation of the tides is an easy subject, if we are contented with a reasonably approximate solution; should, however, [25] it be necessary to discuss fully the tides at any port, the theory of the method necessary for doing so is available, and a most interesting and beautiful theory it certainly is.

Let us then speak for a few moments about the methods by which we can study the tides at a particular port. The principle on which it is based is a very simple one.

It is the month of August, the 18th, we shall suppose, and we are going to enjoy a delicious swim in the sea. We desire, of course, to secure a high tide for the purpose of doing so, and we call an almanac to help us. I refer to the Thom's Dublin Directory, where I find the tide to be high at 10h. 14m. on the morning of the 18th of August. That will then be the time to go down to the baths at Howth or Kingstown.

But what I am now going to discourse to you about is not the delights of sea-bathing, it is rather a different inquiry. I want to ask, How did the people who prepared that almanac know years beforehand, that on that particular day the tide would be high at that particular hour? How do they predict for every day the hour of high water? [26] and how comes it to pass that these predictions are invariably correct?

We first refer to that wonderful book, the *Nautical Almanac*. In that volume the movements of the moon are set forth with full detail; and among other particulars we can learn on page iv of every month the mean time of the moon's meridian passage. It appears that on the day in question the moon crossed the meridian at 11h. 23m. Thus we see there was high water at Dublin at 10h. 14m., and 1h. 9m. later, that is, at 11h. 23m., the moon crossed the meridian.

Let us take another instance. There is a high tide at 3.40 P.M. on the 25th August, and again the infallible *Nautical Almanac* tells us that the moon crossed the meridian at 5h. 44m., that is, at 2h. 4m. after the high water.

In the first case the moon followed the tide in about an hour, and in the second case the moon followed in about two hours. Now if we are to be satisfied with a very rough tide rule for Dublin, we may say generally that there is always a high tide an hour and a half before the moon crosses the meridian. This would not be a very accurate [27] rule, but I can assure you of this, that if you go by it you will never fail of finding a good tide to enable you to enjoy your swim. I do not say this rule would enable you to construct a respectable tide-table. A ship-owner who has to creep up the river, and to whom often the inches of water are material, will require far more accurate tables than this simple rule could give. But we enter into rather complicated matters when we attempt to give any really accurate methods of computation. On these we shall say a few words presently. What I first want to do, is to impress upon you in a simple way the fact of the relation between the tide and the moon.

To give another illustration, let us see how the tides at London Bridge are related to the moon. On Jan. 1st, 1887, it appeared that the tide was high at 6h. 26m. P.M., and that the moon had crossed the meridian 56m. previously; on the 8th Jan. the tide was high at 0h. 43m. P.M., and the moon had crossed the meridian 2h. 1m. previously. Therefore we would have at London Bridge high water following the moon's transit in somewhere about an hour and a half. [28]

I choose a day at random, for example — the 12th April. The moon crosses the upper meridian at 3h. 39m. A.M., and the lower meridian at 4h 6m. P.M. Adding an hour and a half to each would give the high tides at 5h 9m. A.M. and 5h. 36m. P.M.; as a matter of fact, they are 4h. 58m. A.M. and 5h. 20m. P.M.

But these illustrations are sufficient. We find that at London, in a general way, high water appears at London Bridge about an hour and a half after the moon has passed the meridian of London. It so happens that the interval at Dublin is about the same, *i.e.* an hour and a half; only that in the latter case the high water precedes the moon by that interval instead of following it. We may employ the same simple process at other places. Choose two days about a week distant; find on each occasion the interval between the transit of the moon and the time of high water, then the mean of these two differences will always give some notion of the interval between high water and the moon's transit. If then we take from the *Nautical Almanac* the time of the moon's transit, and apply to it the correction proper for the [29] port, we shall always have a sufficiently good tide-table to guide us in choosing a suitable time for taking our swim or our walk by the sea-side; though if you be the captain of a vessel, you will not be so imprudent as to enter port without taking counsel of the accurate tide-tables, for which we are indebted to the Admiralty.

Every one who visits the sea-side, or who lives at a sea-port, should know this constant for the tides, which affect him and his movements so materially. If he will discover it from his own experience, so much the better.

The first point to be ascertained is the time of high water. Do not take this from any local table; you ought to observe it for yourself.

You will go to the pier head, or, better still, to some place where the rise and fall of the mere waves of the sea will not embarrass you in your work. You must note by your watch the time when the tide is highest. An accurate way of doing this will be to have a scale on which you can measure the height at intervals of five minutes about the time of high water. You will then be able to conclude the time at which the tide was actually at its highest [30] point; but even if no great accuracy be obtainable, you can still get much interesting information, for you will without much difficulty be right within ten minutes or a quarter of an hour.

The correction for the port is properly called the "establishment," this being the average time of high water on the days of full and change of the moon at the particular port in question.

We can considerably amend the elementary notion of the tides which the former method has given us, if we adopt the plan described by Dr. Whewell in the first four editions of the *Admiralty Manual of Scientific Inquiry*. We speak of the interval between the transit of the moon and the time of high water as the luni-tidal interval. Of course at full and change this is the same thing as the establishment, but for other phases of the moon the establishment must receive a correction before being used as the luni-tidal interval. The correction is given by the following table—

Hour of Moon's transit after Sun:												
0	1	2	3	4	5	6	7	8	9	10	11
0	-20m	-30m	-50m	-60m	-60m	-60m	-40m	-10m	+10m	+20m	+1
Correction of establishment to find luni-tidal interval:												

Thus at a port where the establishment was 3h. [31] 25m., let us suppose that the transit of the moon took place at 6 P.M.; then we correct the establishment by -60m., and find the luni-tidal interval to be 2h. 25m., and accordingly the high water takes place at 8h. 25m. P.M.

But even this method is only an approximation. The study of the tides is based on accurate observation of their rise and fall on different places round the earth. To show how these observations are to be made, and how they are to be discussed and reduced when they have been made, I may refer to the last edition of the *Admiralty Manual of Scientific Inquiry*, 1886. For a complete study of the tides at any port a self-registering tide-gauge should be erected, on which not alone the heights and times of high and low water should be depicted, but also the continuous curve which shows at any time the height of the water. In fact, the whole subject of the practical observation and discussion and prediction of tides is full of valuable instruction, and may be cited as one of the most complete examples of the modern scientific methods.

In the first place, the tide-gauge itself is a [32] delicate instrument; it is actuated by a float which rises and falls with the water, due provision being made that the mere influence of waves shall not make it to oscillate inconveniently. The motion of the float when suitably reduced by mechanism serves to guide a pencil, which, acting on the paper round a revolving drum, gives a faithful and unintermitting record of the height of the water.

Thus what the tide-gauge does is to present to us a long curved line of which the summits correspond to the heights of high water, while the depressions are the corresponding points of low water. The long undulations of this curve are, however, very irregular. At spring tides, when the sun and the moon conspire, the elevations rise much higher and the depressions sink much lower than they do at neap tides, when the high water raised by the moon is reduced by the action of the sun. There are also many minor irregularities which show the tides to be not nearly such simple phenomena as might be at first supposed. But what we might hastily think of as irregularities are, in truth, the most interesting parts of the whole phenomena. Just as in the observations of the [33] planets the study of the perturbations has led us to results of the widest interest and instruction, so it is these minor phenomena of the tides which seem most pregnant with scientific interest.

The tide-gauge gives us an elaborate curve. How are we to interpret that curve? Here indeed a most beautiful mathematical theorem comes to our aid. Just as ordinary sounds consist of a number of undulations blended together, so the tidal wave consists of a number of distinct undulations superposed. Of these the ordinary lunar tide and the ordinary solar tide are the two principal; but there are also minor undulations, harmonics, so to speak, some originating from the moon, some originating from the sun, and some from both bodies acting in concert.

In the study of sound we can employ an acoustic apparatus for the purpose of decomposing any proposed note, and finding not only the main undulation itself, but the several superposed harmonics which give to the note its timbre. So also we can analyze the undulation of the tide, and show the component parts. The decomposition is effected by the process known as [34] harmonic analysis. The principle of the method may be very simply described. Let us fix our attention on any particular "tide," for so the various elements are denoted. We can always determine beforehand, with as much accuracy as we may require, what the period of that tide will be. For instance, the period of the lunar semi-diurnal tide will of course be half the average time occupied by the moon to travel round from the meridian of any place until it regains the same meridian; the period of the lunar diurnal tide will be double as great; and there are fortnightly tides, and others of periods still greater. The essential point to notice is, that the periods of these tides are given by purely astronomical considerations from the periods of the motions theory, and do not depend upon the actual observations.

We measure off on the curve the height of the tide at intervals of an hour. The larger the number of such measures that are available the better; but even if there be only three hundred and sixty or seven hundred and twenty consecutive hours, then, as shown by Professor G. H. Darwin in the *Admiralty Manual* already referred [35] to, it will still be possible to obtain a very competent knowledge of the tides in the particular port where the gauge has been placed.

The art (for such indeed it may be described) of harmonic analysis consists in deducing from the hourly observations the facts with regard to each of the constituent tides. This art has been carried to such perfection, that it has been reduced to a very simple series of arithmetical operations. Indeed it has now been found possible to call in the aid of ingenious mechanism, by which the labours of computation are entirely superseded. The pointer of the harmonic analyzer has merely to be traced over the curve which the tide-gauge has drawn, and it is the function of the machine to decompose the composite undulation into its parts, and to exhibit the several constituent tides whose confluence gives the total result.

As if nothing should be left to complete the perfection of a process which, both from its theoretical and its practical sides, is of such importance, a machine for predicting tides has been designed, constructed, and is now in ordinary use. When by the aid of the harmonic analysis the effectiveness [36] of the several constituent tides affecting a port have become fully determined, it is of course possible to predict the tides for that port. Each "tide" is a simple periodic rise and fall, and we can compute for any future time the height of each were it acting alone. These heights can all be added together, and thus the height of the water is obtained. In this way a tide-table is formed, and such a table when complete will express not alone the hours and heights of high water on every day, but the height of the water at any intervening hour.

The computations necessary for this purpose are no doubt simple, so far as their principle is concerned; but they are exceedingly tedious, and any process must be welcomed which affords mitigation of a task so laborious. The entire theory of the tides owes much to Sir William Thomson in the methods of observation and in the methods of reduction. He has now completed the practical parts of the subject by inventing and constructing the famous tide-predicting engine.

The principle of this engine is comparatively simple. There is a chain which at one end is [37] fixed, and at the other end carries the pencil which is pressed against the revolving drum on which the prediction is to be inscribed. Between its two ends the chain passes up and down over pulleys. Each pulley corresponds to one of the "tides," and there are about a dozen altogether, some of which exercise but little effect. Of course if the centres of the pulleys were all fixed the pen could not move, but the centre of each pulley describes a circle with a radius proportional to the amplitude of the corresponding tide, and in a time proportional to the period of that tide. When these pulleys are all set so as to start at the proper phases, the motion is produced by turning round a handle which makes the drum rotate, and sets all the pulleys in motion. The tide curve is thus rapidly drawn out; and so expeditious is the machine, that the tides of a port for an entire year can be completely worked out in a couple of hours.

While the student or the philosopher who seeks to render any account of the tide on dynamical grounds is greatly embarrassed by the difficulties introduced by friction, we, for our present purpose [38] in the study of the great romance of modern science opened up to us by the theory of the tides, have to welcome friction as the agent which gives to the tides their significance from our point of view.

There is the greatest difference between the height of the rise and fall of the tide at different localities. Out in mid-ocean, for instance, an island like St. Helena is washed by a tide only about three feet in range; an enclosed sea like the Caspian is subject to no appreciable tides whatever, while the Mediterranean, notwithstanding its connection with the Atlantic, is still only subject to very inconsiderable tides, varying from one foot to a few feet. The statement that water always finds its own level must be received, like many another proposition in nature, with a considerable degree of qualification. Long ere one tide could have found its way through the Straits of Gibraltar in sufficient volume to have appreciably affected the level of the great inland sea, its effects would have been obliterated by succeeding tides. On the other hand, there are certain localities which expose a funnel-shape opening to the sea; into these the great tidal [39] wave rushes, and as it passes onwards towards the narrow part, the waters become piled up so as to produce tidal phenomena of abnormal proportions. Thus, in our own islands, we have in the Bristol Channel a wide mouth into which a great tide enters, and as it hurries up the Severn it produces the extraordinary phenomenon of the Bore. The Bristol Channel also concentrates the great wave which gives Chepstow and Cardiff a tidal range of thirty-seven or thirty-eight feet at springs, and forces the sea up the river Avon so as to give Bristol a wonderful tide. There is hardly any more interesting spot in our islands for the observation of tides than is found on Clifton Suspension Bridge. From that beautiful structure you look down on a poor and not very attractive stream, which two hours later becomes transformed into a river of ample volume, down which great ships are navigated. But of all places in the world, the most colossal tidal phenomena are those in the Bay of Fundy. Here the Atlantic passes into a long channel whose sides gradually converge. When the great pulse of the tide rushes up this channel, it is gradually accumulated into a mighty volume [40] at the upper end, the ebb and flow of which at spring tides extends through the astonishing range of not less than fifty feet.

These discrepancies between the tides at different places are chiefly due to the local formations of the coasts and the sea-beds. Indeed, it seems that if the whole earth were covered with an uniform and deep ocean of water, the tides would be excessively feeble. On no other supposition can we reasonably account for the fact that our barometric records fail to afford us any very distinct evidence as to the existence of tides in the atmosphere. For you will, of course, remember that our atmosphere may be regarded as a deep and vast ocean of air, which embraces the whole earth, extending far above the loftiest summits of the mountains.

It is one of the profoundest of nature's laws that wherever friction takes place, energy has to be consumed. Perhaps I ought rather to say transformed, for of course it is now well known that consumption of energy in the sense of absolute loss is impossible. Thus, when energy is expended in moving a body in opposition to the force of [41] friction, or in agitating a liquid, the energy which disappears in its mechanical form reappears in the form of heat. The agitation of water by paddles moving through it warms the water, and the accession of heat thus acquired measures the energy which has been expended in making the paddles rotate. The motion of a liquid of which the particles move among each other with friction, can only be sustained by the incessant degradation of energy from the mechanical form into the lower form of diffused heat. Thus the very fact that the tides are ebbing and flowing, and that there is consequently incessant friction going on among all the particles of water in the ocean, shows us that there must be some great store of energy constantly available to supply the incessant draughts made upon it by the daily oscillation of the tides. In addition to the mere friction between the particles of water, there are also many other ways in which the tides proclaim to us that there is some great hoard of energy which is continually accessible to their wants. Stand on the bank of an estuary or river up and down which a great tidal current ebbs and flows; you [42] will see the water copiously charged with sediment which the tide is bearing along. Engineers are well aware of the potency of the tide as a vehicle for transporting stupendous quantities of sand or mud. A sand-bank impedes the navigation of a river; the removal of that sand-bank would be a task, perhaps, conceivably possible by the use of steam dredges and other appliances, whereby vast quantities of sand could be raised and transported to another locality where they would be innocuous. It is sometimes possible to effect the desired end by applying the power of the tide. A sea-wall judiciously thrown out will sometimes concentrate the tide into a much narrower channel. Its daily oscillations will be accomplished with greater vehemence, and as the tide rushes furiously backwards and forwards over the obstacle, the incessant action will gradually remove it, and the impediment to navigation may be cleared away. Here we actually see the tides performing a piece of definite and very laborious work, to accomplish which by the more ordinary agents would be a stupendous task.

In some places the tides are actually harnessed so as to accomplish useful work. I have read that [43] underneath old London Bridge there used formerly to be great water-wheels, which were turned by the tide as it rushed up the river, and turned again, though in the opposite way, by the ebbing tide. These wheels were, I believe, employed to pump up water, though it does not seem obvious for what purposes the water would have been suitable. Indeed in the ebb and flow all round our coasts there is a potential source of energy which has hitherto been allowed to run to waste. The tide could be utilized in various ways. Many of you will remember the floating mills on the Rhine. They are vessels like paddle steamers anchored in the rapid current. The flow of the river makes the paddles rotate, and thus the machinery in the interior is worked. Such craft moored in a rapid tide-way could also be made to convey the power of the tides into the mechanism of the mill. Or there is still another method which has been employed, and which will perhaps have a future before it in those approaching times when the coal-cellars of England shall be exhausted. Imagine on the sea-coast a large flat extent which is inundated twice every day by the tide. Let us build a stout [44] wall round this area, and provide it with a sluice-gate. Open the gate as the tide rises, and the great pond will be filled; then at the moment of high water close the sluice, and the pond-full will be impounded. If at low tide the sluice be opened the water will rush tumultuously out. Now suppose that a water-wheel be provided, so that the rapid rush of water from the exit shall fall upon its blades; then a source of power is obviously the result.

At present, however, such a contrivance would naturally find no advocates, for of course the commercial aspect of the question is that which will decide whether the scheme is practicable and economical. The issue indeed can be very simply stated. Suppose that a given quantity of power be required — let us say that of one hundred horse. Then we have to consider the conditions under which a contrivance of the kind we have sketched shall yield a power of this amount. Sir William Thomson, in a very interesting address to the British Association at York in 1881, discussed this question, and I shall here make use of the facts he brought forward on that occasion. He showed that to [45] obtain as much power as could be produced by a steam-engine of one hundred horse power, a very large reservoir would be required. It is doubtful indeed whether there would be many localities on the earth which would be suitable for the purpose. Suppose, however, an estuary could be found which had an area of forty acres; then if a wall were thrown across the mouth so that the tide could be impounded, the total amount of power that could be yielded by a water-wheel worked by the incessant influx and efflux of the tide would be equal to that yielded by the one hundred horse engine, running continuously from one end of the year to the other.

There are many drawbacks to a tide-mill of this description. In the first place, its situation would naturally be far removed from other conveniences necessary for manufacturing purposes. Then too there is the great irregularity in the way in which the power is rendered available. At certain periods during the twenty-four hours the mill would stop running, and the hours when this happened would be constantly changing. The inconvenience from the manufacturer's point of [46] view of a deficiency of power during neap tides might not be compensated by the fact that he had an excessive supply of power at spring-tides. Before tide-mills could be suitable for manufacturing purposes, some means must be found for storing away the energy when it is redundant, and applying it when its presence is required. We should want in fact for great sources of energy some contrivance which shall fulfil the same purpose as the accumulators do in an electrical installation.

Even then, however, the financial consideration remains, as to whether the cost of building the dam and maintaining the tide-mill in good order will not on the whole exceed the original price and the charges for the maintenance of a hundred horse power steam-engine. There cannot be a doubt that in this epoch of the earth's history, so long as the price of coal is only a few shillings a ton, the tide-mill, even though we seem to get its power without current expense, is vastly more expensive than a steam-engine. Indeed, Sir William Thomson remarks, that wherever a suitable tidal basin could be found, it would be nearly as easy to reclaim the land altogether from the sea. And if [47] this were in any locality where manufactures were possible, the commercial value of forty acres of reclaimed land would greatly exceed all the expenses attending the steam-engine. But when the time comes, as come it apparently will, that the price of coal shall have risen to several pounds a ton, the economical aspect of steam as compared with other prime movers will be greatly altered; it will then no doubt be found advantageous to utilize great sources of energy, such as Niagara and the tides, which it is now more prudent to let run to waste.

For my argument, however, it matters little that the tides are not constrained to do much useful work. They are always doing work of some kind, whether that be merely heating the particles of water by friction, or vaguely transporting sand from one part of the ocean to the other. Useful work or useless work are alike for the purpose of my argument. We know that work can never be done unless by the consumption or transformation of energy. For each unit of work that is done—whether by any machine or contrivance, by the muscles of man or any other animal, by the winds, the waves, or the tides, or in any other way [48] whatever—a certain equivalent quantity of energy must have been expended. When, therefore, we see any work being performed, we may always look for the source of energy to which the machine owes its efficiency. In fact, it is the old story illustrated, that perpetual motion is impossible. A mechanical device, however ingenious may be the construction, or however accurate the workmanship, can never possess what is called perpetual motion. It is needless to enter into details of any proposed contrivance of wheels, of pumps, of pulleys; it is sufficient to say that nothing in the shape of mechanism can work without friction, that friction produces heat, that heat is a form of energy, and that to replace the energy consumed in producing the heat there must be some source from which the machine is replenished if its motion is to be continued indefinitely.

Hence, as the tides may be regarded as a machine doing work, we have to ascertain the origin of that energy which they are continually expending. It is at this point that we first begin to feel the difficulties inherent in the theory of tidal evolution. I do not mean difficulties in the sense of doubts, for [49] up to the present I have mentioned no doubtful point. When I come to such I shall give due warning. By difficulties I now mean points which it is not easy to understand without a little dynamical theory; but we must face these difficulties, and endeavour to elucidate them as well as we can.

Let us first see what the sources of energy can possibly be on which the tides are permitted to draw. Our course is simplified by the fact that the energy of which we have to speak is of a mechanical description, that is to say, not involving heat or other more obscure forms of energy. A simple type of energy is that possessed by a clock-weight after the clock has been wound. A store of power is thus laid up which is gradually doled out during the week in small quantities, second by second, to sustain the motion of the pendulum. The energy in this case is due to the fact that the weight is attracted by the earth, and is yielded according as the weight sinks downwards. In the separation between two mutually attracting bodies, a store of energy is thus implied. What we learn from an ordinary [50] clock may be extended to the great bodies of the universe. The moon is a gigantic globe separated from our earth by a distance of 240,000 miles. The attraction between these two bodies always tends to bring them together. No doubt the moon is not falling towards the earth as the descending clock-weight is doing.

We may, in fact, consider the moon, so far as our present object is concerned, to be revolving almost in a circle, of which the earth is the centre. If the moon, however, were to be stopped, it would at once commence to rush down towards the earth, whither it would arrive with an awful crash in the course of four or five days. It is fortunately true that the moon does not behave thus; but it has the ability of doing so, and thus the mere separation between the earth and the moon involves the existence of a stupendous quantity of energy, capable under certain conditions of undergoing transformation.

There is also another source of mechanical energy besides that we have just referred to. A rapidly moving body possesses, in virtue of its motion, a store of readily available energy, and it is easy to show that energy of this type is capable of transformation [51] into other types. Think of a cannon-ball rushing through the air at a speed of a thousand feet per second; it is capable of wreaking disaster on anything which it meets, simply because its rapid motion is the vehicle by which the energy of the gunpowder is transferred from the gun to where the blow is to be struck. Had the cannon been directed vertically upwards, then the projectile, leaving the muzzle with the same initial velocity as before, would soar up and up, with gradually abating speed, until at last it reached a turning-point, the elevation of which would depend upon the initial velocity. Poised for a moment at the summit, the cannon-ball may then be likened to the clock-weight, for the entire energy which it possessed by its motion has been transformed into the statical energy of a raised weight. Thus we see these two forms of energy are mutually interchangeable. The raised weight if allowed to fall will acquire velocity, or the rapidly moving weight if directed upwards will acquire altitude.

The quantity of energy which can be conveyed by a rapidly moving body increases greatly with its speed. For instance, if the speed of the body [52] be doubled, the energy will be increased fourfold, or, in general, the energy which a moving body possesses may be said to be proportional to the square of its speed. Here then we have another source of the energy present in our earth-moon system; for the moon is hurrying along in its path with a speed of two-thirds of a mile per second, or about twice or three times the speed of a cannon-shot. Hence the fact that the moon is continuously revolving in a circle shows us that it possesses a store of energy which is nine times as great as that which a cannon-ball as massive as the moon, and fired with the ordinary velocity, would receive from the powder which discharged it.

Thus we see that the moon is endowed with two sources of energy, one of which is due to its separation from the earth, and the other to the speed of its motion. Though these are distinct, they are connected together by a link which it is important for us to comprehend. The speed with which the moon revolves around the earth is connected with the moon's distance from the earth. The moon might, for instance, revolve in a larger circle than that which it actually pursues; but if [53] it did so, the speed of its motion would have to be appropriately lessened. The orbit of the moon might have a much smaller radius than it has at present, provided that the speed was sufficiently increased to compensate for the increased attraction which the earth would exercise at the lessened distance. Indeed, I am here only stating what every one is familiar with under the form of Kepler's Law, that the square of the periodic time is in proportion to the cube of the mean distance. To each distance of the moon therefore belongs an appropriate speed. The energy due to the moon's position and the energy due to its motion are therefore connected together. One of these quantities cannot be altered without the other undergoing change. If the moon's orbit were increased there would be a gain of energy due to the enlarged distance, and a loss of energy due to the diminished speed. These would not, however, exactly compensate. On the whole, we may represent the total energy of the moon as a single quantity, which increases when the distance of the moon from the earth increases, and lessens when the distance from the earth to the moon [54] lessens. For simplicity we may speak of this as moon-energy.

But the most important constituent of the store of energy in the earth-moon system is that contributed by the earth itself. I do not now speak of the energy due to the velocity of the earth in its orbit round the sun. The moon indeed participates in this equally with the earth, but it does not affect those mutual actions between the earth and moon with which we are at present concerned. We are, in fact, discussing the action of that piece of machinery the earth-moon system; and its action is not affected by the circumstance that the entire machine is being bodily transported around the sun in a great annual voyage. This has little more to do with the action of our present argument than has the fact that a man is walking about to do with the motions of the works of the watch in his pocket. We shall, however, have to allude to this subject further on.

The energy of the earth which is significant in the earth-moon theory is due to the earth's rotation upon its axis. We may here again use as an illustration the action of machinery; and the [55] special contrivance that I now refer to is the punching-engine that is used in our ship-building works. In preparing a plate of iron to be riveted to the side of a ship, a number of holes have to be made all round the margin of the plate. These holes must be half an inch or more in diameter, and the plate is sometimes as much as, or more than, half an inch in thickness. The holes are produced in the metal by forcing a steel punch through it; and this is accomplished without even heating the plate so as to soften the iron. It is needless to say that an intense force must be applied to the punch. On the other hand, the distance through which the punch has to be moved is comparatively small. The punch is attached to the end of a powerful lever, the other end of the lever is raised by a cam, so as to depress the punch to do its work. An essential part of the machine is a small but heavy fly-wheel connected by gearing with the cam.

This fly-wheel when rapidly revolving contains within it, in virtue of its motion, a large store of energy which has gradually accumulated during the time that the punch is not actually in action. [56] The energy is no doubt originally supplied from a steam-engine. What we are especially concerned with is the action of the rapidly rotating wheel as a reservoir in which a large store of energy can be conveniently maintained until such time as it is wanted. In the action of punching, when the steel die comes down upon the surface of the plate, a large quantity of energy is suddenly demanded to force the punch against the intense resistance it experiences; the energy for this purpose is drawn from the store in the fly-wheel, which experiences no doubt a check in its velocity, to be regained again from the energy of the engine during the interval which elapses before the punch is called on to make the next hole.

Another illustration of the fly-wheel on a splendid scale is seen in our mighty steel works, where ponderous rails are being manufactured. A white-hot ingot of steel is presented to a pair of powerful rollers, which grip the steel, and send it through at the other side both compressed and elongated. Tremendous power is required to meet the sudden demand on the machine at the critical moment. To obtain this power an engine of stupendous [57] proportions is sometimes attached directly to the rollers, but more frequently an engine of rather less horse-power will be used, the might of this engine being applied to giving rapid rotation to an immense fly-wheel, which may thus be regarded as a reservoir full of energy. The rolling mills then obtain from this store in the fly-wheel whatever energy is necessary for their gigantic task.

These illustrations will suffice to show how a rapidly rotating body may contain energy in virtue of its rotation, just as a cannon-ball contains energy in virtue of its speed of translation, or as a clock-weight has energy in virtue of the fact that it has some distance to fall before it reaches the earth. The rotating body need not necessarily have the shape of a wheel—it may be globular in form; nor need the axes of rotation be fixed in bearings, like those of the fly-wheel; nor of course is there any limit to the dimensions which the rotating body may assume. Our earth is, in fact, a vast rotating body 8000 miles in diameter, and turning round upon its axis once every twenty-three hours and fifty-six minutes. Viewed in this way, the earth is to be regarded as a gigantic [58] fly-wheel containing a quantity of energy great in correspondence with the earth's mass. The amount of energy which can be stored by rotation also depends upon the square of the velocity with which the body turns round; thus if our earth turned round in half the time which it does at present, that is, if the day was twelve hours instead of twenty-four hours, the energy contained in virtue of that rotation would be four times its present amount.

Reverting now to the earth-moon system, the energy which that system contains consists essentially of two parts—the moon-energy, whose composite character I have already explained, and the earth-energy, which has its origin solely in the rotation of the earth on its axis. It is necessary to observe that these are essentially distinct—there is no necessary relation between the speed of the earth's rotation and the distance of the moon, such as there is between the distance of the moon and the speed with which it revolves in its orbit.

For completeness, it ought to be added that there is also some energy due to the moon's rotation on its axis, but this is very small for two [59] reasons: first, because the moon is small compared with the earth, and second, because the angular velocity of the moon is also very small compared with that of the earth. We may therefore dismiss as insignificant the contributions from this source of energy to the sum total.

I have frequently used illustrations derived from machinery, but I want now to emphasize the profound distinction that exists between the rotation of the earth and the rotation of a fly-wheel in a machine shop. They are both, no doubt, energy-holders, but it must be borne in mind, that as the fly-wheel doles out its energy to supply the wants of the machines with which it is connected, a restitution of its store is continually going on by the action of the engine, so that on the whole the speed of the fly-wheel does not slacken. The earth, however, must be likened to a fly-wheel which has been disconnected with the engine. If, therefore, the earth have to supply certain demands on its accumulation of energy, it can only do so by a diminution of its hoard, and this involves a sacrifice of some of its speed.

In the earth-moon system there is no engine [60] at hand to restore the losses of energy which are inevitable when work has to be done. But we have seen that work is done; we have shown, in fact, that the tides are at present doing work, and have been doing work for as long a period in the past as our imagination can extend to. The energy which this work has necessitated can only have been drawn from the existing store in the system; that energy consists of two parts—the moon-energy and the earth's rotation energy. The problem therefore for us to consider is, which of these two banks the tides have drawn on to meet their constant expenditure. This is not a question that can be decided offhand; in fact, if we attempt to decide it in an offhand manner we shall certainly go wrong. It seems so very plausible to say that as the moon causes the tides, therefore the energy which these tides expend should be contributed by the moon. But this is not the case. It actually happens that though the moon does cause the tides, yet when those tides consume energy they draw it not from the distant moon, but from the vast supply which they find ready to their hand, stored up in the rotation of the earth. [61]

The demonstration of this is not a very simple matter; in fact, it is so far from being simple that many philosophers, including some eminent ones too, while admitting that of course the tides must have drawn their energy from one or other or both of these two sources, yet found themselves unable to assign how the demand was distributed between the two conceivable sources of supply.

We are indebted to Professor Purser of Belfast for having indicated the true dynamical principle on which the problem depends. It involves reasoning based simply on the laws of motion and on elementary mathematics, but not in the least involving questions of astronomical observation. It would be impossible for me in a lecture like this to give any explanation of the mathematical principles referred to. I shall, however, endeavour by some illustrations to set before you what this profound principle really is. Were I to give it the old name I should call it the law of the conservation of areas; the more modern writers, however, speak of it as the conservation of moment of momentum, an expression which exhibits the nature of the principle in a more definite manner. [62]

I do not see how to give any very accurate illustration of what this law means, but I must make the attempt, and if you think the illustration beneath the dignity of the subject, I can only plead the difficulty of mathematics as an excuse. Let us suppose that a ballroom is fairly filled with dancers, or those willing to dance, and that a merry waltz is being played; the couples have formed, and the floor is occupied with pairs who are whirling round and round in that delightful amusement. Some couples drop out for a while and others strike in; the fewer couples there are the wider is the range around which they can waltz, the more numerous the couples the less individual range will they possess. I want you to realize that in the progress of the dance there is a certain total quantity of spin at any moment in progress; this spin is partly made up of the rotation by which each dancer revolves round his partner, and partly of the circular orbit about the room which each couple endeavours to describe. If there are too many couples on the floor for the general enjoyment of the dance, then both the orbit and the angular velocity of each couple will be [63] restricted by the interference with their neighbours. We may, however, assert that so long as the dance is in full swing the total quantity of spin, partly rotational and partly orbital, will remain constant. When there are but few couples the unimpeded rotation and the large orbits will produce as much spin as when there is a much larger number of couples, for in the latter case the diminished freedom will lessen the quantity of spin produced by each individual pair. It will sometimes happen too that collision will take place, but the slight diversions thus arising only increase the general merriment, so that the total quantity of spin may be sustained, even though one or two couples are placed temporarily *hors de combat*. I have invoked a ball-room for the purpose of bringing out what we may call the law of the conservation of spin. No matter how much the individual performers may change, or no matter what vicissitudes arise from their collision and other mutual actions, yet the total quantity of spin remains unchanged.

Let us look at the earth-moon system. The law of the conservation of moment of momentum [64] may, with sufficient accuracy for our present purpose, be interpreted to mean that the total quantity of spin in the system remains unaltered. In our system the spin is threefold; there is first the rotation of the earth on its axis, there is the rotation of the moon on its axis, and then there is the orbital revolution of the moon around the earth. The law to which we refer asserts that the total quantity of these three spins, each estimated in the proper way, will remain constant. It matters not that tides may ebb and flow, or that the distribution of the spin shall vary, but its total amount remains inflexibly constant. One constituent of the total amount — that is, the rotation of the moon on its axis — is so insignificant, that for our present purposes it may be entirely disregarded. We may therefore assert that the amount of spin in the earth, due to its rotation round its axis, added to the amount of spin in the moon due to its revolution round the earth, remains unalterable. If one of these quantities change by increase or by decrease, the other must correspondingly change by decrease or by increase. If, therefore, from any cause, the earth began to spin a little [65] more quickly round its axis, the moon must do a little less spin; and consequently, it must shorten its distance from the earth. Or suppose that the earth's velocity of rotation is abated, then its contribution to the total amount of spin is lessened; the deficiency must therefore be made up by the moon, but this can only be done by an enlargement of the moon's orbit. I should add, as a caution, that these results are true only on the supposition that the earth-moon system is isolated from all external interference. With this proviso, however, it matters not what may happen to the earth or moon, or what influence one of them may exert upon the other, no matter what tides may be raised, no matter even if the earth fly into fragments, the whole quantity of spin of all those fragments would, if added to the spin of the moon, yield the same unalterable total. We are here in possession of a most valuable dynamical principle. We are not concerned with any special theory as to the action of the tides; it is sufficient for us that in some way or other the tides have been caused by the moon, and that being so, the principle of the conservation of spin will apply. [66]

Were the earth and the moon both rigid bodies, then there could be of course no tides on the earth, it being rigid and devoid of ocean. The rotation of the earth on its axis would therefore be absolutely without change, and therefore the necessary condition of the conservation of spin would be very simply attained by the fact that neither of the constituent parts changed. The earth, however, not being entirely rigid, and being subject to tides, this simple state of things cannot continue; there must be some change in progress.

I have already shown that the fact of the ebbing and the flowing of the tide necessitates an expenditure of energy, and we saw that this energy must come either from that stored up in the earth by its rotation, or from that possessed by the moon in virtue of its distance and revolution. The law of the conservation of spin will enable us to decide at once as to whence the tides get their energy. Suppose they took it from the moon, the moon would then lose in energy, and consequently come nearer the earth. The quantity of spin contributed by the moon would therefore be lessened, and accordingly the spin to be made up by the [67] earth would be increased. That means, of course, that the velocity of the earth rotating on its axis must be increased, and this again would necessitate an increase in the earth's rotational energy. It can be shown, too, that to keep the total spin right, the energy of the earth would have to gain more than the moon would have lost by revolving in a smaller orbit. Thus we find that the total quantity of energy in the system would be increased. This would lead to the absurd result that the action of the tides manufactured energy in our system. Of course, such a doctrine cannot be true; it would amount to a perpetual motion! We might as well try to get a steam-engine which would produce enough heat by friction not only to supply its own boilers, but to satisfy all the thermal wants of the whole parish. We must therefore adopt the other alternative. The tides do not draw their energy from the moon; they draw it from the store possessed by the earth in virtue of its rotation.

We can now state the end of this rather long discussion in a very simple and brief manner. Energy can only be yielded by the earth at the [68] expense of some of the speed of its rotation. The tides must therefore cause the earth to revolve more slowly; in other words, *the tides are increasing the length of the day.*

The earth therefore loses some of its velocity of rotation; consequently it does less than its due share of the total quantity of spin, and an increased quantity of spin must therefore be accomplished by the moon; but this can only be done by an enlargement of its orbit. Thus there are two great consequences of the tides in the earth-moon system—the days are getting longer, the moon is receding further.

These points are so important that I shall try and illustrate them in another way, which will show, at all events, that one and both of these tidal phenomena commend themselves to our common sense. Have we not shown how the tides in their ebb and flow are incessantly producing friction, and have we not also likened the earth to a great wheel? When the driver wants to stop a railway train the brakes are put on, and the brake is merely a contrivance for applying friction to the circumference of a wheel for the purpose of checking its [69] motion. Or when a great weight is being lowered by a crane, the motion is checked by a band which applies friction on the circumference of a wheel, arranged for the special purpose. Need we then be surprised that the friction of the tides acts like a brake on the earth, and gradually tends to check its mighty rotation? The progress of lengthening the day by the tides is thus readily intelligible. It is not quite so easy to see why the ebbing and the flowing of the tide on the earth should actually have the effect of making the moon to retreat; this phenomenon is in deference to a profound law of nature, which tells us that action and reaction are equal and opposite to each other. If I might venture on a very homely illustration, I may say that the moon, like a troublesome fellow, is constantly annoying the earth by dragging its waters backward and forward by means of tides; and the earth, to free itself from this irritating interference, tries to push off the aggressor and to make him move further away.

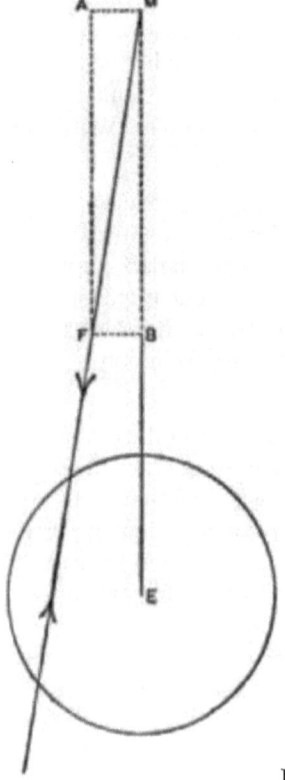

Fig. 2.

Another way in which we can illustrate the retreat of the moon as the inevitable consequence of tidal friction is shown in the adjoining figure, in [70] which the large body E represents the earth, and the small body M the moon. We may for simplicity regard the moon as a point, and as this attracts each particle of the earth, the total effect of the moon on the earth may be represented by a single force. By the law of equality of action and reaction, the force of the earth on the moon is to be represented by an equal and opposite force. If there were no tides then the moon's force would of course pass through the earth's centre; but as the effect of the moon is to slacken the earth's rotation, it follows that the total force does not exactly pass through the line of the [71] earth's centre, but a little to one side, in order to pull the opposite way to that in which the earth is turning, and thus bring down its speed. We may therefore decompose the earth's total force on the moon into two parts, one of which tends directly towards the earth's centre, while the other acts tangentially to the moon's orbit. The central force is of course the main guiding power which keeps the moon in its path; but the incessant tangential force constantly tends to send the moon out further and further, and thus the growth of its orbit can be accounted for.

We therefore conclude finally, that the tides are making the day longer and sending the moon away further. It is the development of the consequences of these laws that specially demands our attention in these lectures. We must have the courage to look at the facts unflinchingly, and deduce from them all the wondrous consequences they involve. Their potency arises from a characteristic feature — they are unintermitting. Most of the great astronomical changes with which we are ordinarily familiar are really periodic: they gradually increase in one direction for years, for centuries, or [72] for untold ages; but then a change comes, and the increase is changed into a decrease, so that after the lapse of becoming periods the original state of things is restored. Such periodic phenomena abound in astronomy. There is the annual fluctuation of the seasons; there is the eighteen or nineteen year period of the moon; there is the great period of the precession of the equinoxes, amounting to twenty-six thousand years; and then there is the stupendous Annus Magnus of hundreds of thousands of years, during which the earth's orbit itself breathes in and out in response to the attraction of the planets. But these periodic phenomena, however important they may be to us mere creatures of a day, are insignificant in their effects on the grand evolution through which the celestial bodies are passing. The really potent agents in fashioning the universe are those which, however slow or feeble they may seem to be, are still incessant in their action. The effect which a cause shall be competent to produce depends not alone upon the intensity of that cause, but also upon the time during which it has been in operation. From the phenomena of geology, as well as from [73] those of astronomy, we know that this earth and the system to which it belongs has endured for ages, not to be counted by scores of thousands of years, or, as Prof. Tyndall has so well remarked, "Not for six thousand years, nor for sixty thousand years, nor six hundred thousand years, but for æons of untold millions." Those slender agents which have devoted themselves unceasingly to the accomplishment of a single task may in this long lapse of time have accomplished results of stupendous magnitude. In famed stalactite caverns we are shown a colossal figure of crystal extending from floor to roof, and the formation of that column is accounted for when we see a tiny drop falling from the roof above to the floor beneath. A lifetime may not suffice for that falling drop to add an appreciable increase to the stalactite down which it trickles, or to the growing stalagmite on which it falls; but when the operation has been in progress for immense ages, it is capable of the formation of the stately column. Here we have an illustration of an influence which, though apparently trivial, acquires colossal significance when adequate time is afforded. It is phenomena of this kind [74] which the student of nature should most narrowly watch, for they are the real architects of the universe.

The tidal consequences which we have already demonstrated are emphatically of this non-periodic class—the day is always lengthening, the moon is always retreating. To-day is longer than yesterday; to-morrow will be longer than to-day. It cannot be said that the change is a great one; it is indeed too small to be appreciable even by our most delicate observations. In one thousand years the alteration in the length of a day is only a small fraction of a second; but what may be a very small matter in one thousand years can become a very large one in many millions of years. Thus it is that when we stretch our view through immense vistas of time past, or when we look forward through immeasurable ages of time to come, the alteration in the length of the day will assume the most startling proportions, and involve the most momentous consequences.

Let us first look back. There was a time when the day, instead of being the twenty-four hours we now have, must have been only twenty-three hours, [75] How many millions of years ago that was I do not pretend to say, nor is the point material for our argument; suffice it to say, that assuming, as geology assures us we may assume, the existence of these æons of millions of years, there was once a time when the day was not only one hour shorter, but was even several hours less than it is at present. Nor need we stop our retrospect at a day of even twenty, or fifteen, or ten hours long; we shall at once project our glance back to an immeasurably remote epoch, at which the earth was spinning round in a time only one sixth or even less of the length of the present day. There is here a reason for our retrospect to halt, for at some eventful period, when the day was about three or four hours long, the earth must have been in a condition of a very critical kind.

It is well known that fearful accidents occasionally happen where large grindstones are being driven at a high speed. The velocity of rotation becomes too great for the tenacity of the stone to withstand the stress; a rupture takes place, the stone flies in pieces, and huge fragments are hurled around. For each particular grindstone [76] there is a certain special velocity depending upon its actual materials and character, at which it would inevitably fly in pieces. I have once before likened our earth to a wheel; now let me liken it to a grindstone. There is therefore a certain critical velocity of rotation for the earth at which it would be on the brink of rupture. We cannot exactly say, in our ignorance of the internal constitution of the earth, what length of day would be the shortest possible for our earth to have consistently with the preservation of its integrity; we may, however, assume that it will be about three or four hours, or perhaps a little less than three. The exact amount, however, is not really very material to us; it would be sufficient for our argument to assert that there is a certain minimum length of day for which the earth can hold together. In our retrospect, therefore, through the abyss of time past our view must be bounded by that state of the earth when it is revolving in this critical period. With what happened before that we shall not at present concern ourselves. Thus we look back to a time at the beginning of the present order of things, when the day was only some three or four hours long. [77]

Let us now look at the moon, and examine where it must have been during these past ages. As the moon is gradually getting further and further from us at present, so, looking back into past time, we find that the moon was nearer and nearer to the earth the further back our view extends; in fact, concentrating our attention solely on essential features, we may say that the path of the moon is a sort of spiral which winds round and round the earth, gradually getting larger, though with extreme slowness. Looking back we see this spiral gradually coiling in and in, until in a retrospect of millions of years, instead of its distance from the earth being 240,000 miles, it must have been much less. There was a time when the moon was only 200,000 miles away; there was a time many millions of years ago, when the moon was only 100,000 miles away. Nor can we here stop our retrospect; we must look further and further back, and follow the moon's spiral path as it creeps in and in towards the earth, until at last it appears actually in contact with that great globe of ours, from which it is now separated by a quarter of a million of miles. [78]

Surely the tides have thus led us to the knowledge of an astounding epoch in our earth's past history, when the earth is spinning round in a few hours, and when the moon is, practically speaking, in contact with it. Perhaps I should rather say, that the materials of our present moon were in this situation, for we would hardly be entitled to assume that the moon then possessed the same globular form in which we see it now. To form a just apprehension of the true nature of both bodies at this critical epoch, we must study their concurrent history as it is disclosed to us by a totally different line of reasoning.

Drop, then, for a moment all thought of tides, and let us bring to our aid the laws of heat, which will disclose certain facts in the ancient history of the earth-moon system perhaps as astounding as those to which the tides have conducted us. In one respect we may compare these laws of heat with the laws of the tides; they are both alike non-periodic, their effects are cumulative from age to age, and imagination can hardly even impose a limit to the magnificence of the works they can accomplish. Our argument from heat is founded [79] on a very simple matter. It is quite obvious that a heated body tends to grow cold. I am not now speaking of fires or of actual combustion whereby heat is produced; I am speaking merely of such heat as would be possessed by a red-hot poker after being taken from the fire, or by an iron casting after the metal has been run into the mould. In such cases as this the general law holds good, that the heated body tends to grow cold. The cooling may be retarded no doubt if the passage of heat from the body is impeded. We can, for instance, retard the cooling of a teapot by the well-known practice of putting a cosy upon it; but the law remains that, slowly or quickly, the heated body will tend to grow colder. It seems almost puerile to insist with any emphasis on a point so obvious as this, but yet I frequently find that people do not readily apprehend all the gigantic consequences that can flow from a principle so simple. It is true that a poker cools when taken from the fire; we also find that a gigantic casting weighing many tons will grow gradually cold, though it may require days to do so. The same principle will extend to any object, no matter how [80] vast it may happen to be. Were that great casting 2000 miles in diameter, or were it 8000 miles in diameter, it will still steadily part with its heat, though no doubt the process of cooling becomes greatly prolonged with an increase in the dimensions of the heated body. The earth and the moon cannot escape from the application of these simple principles.

Let us first speak of the earth. There are multitudes of volcanoes in action at the present moment in various countries upon this earth. Now whatever explanation may be given of the approximate cause of the volcanic phenomena, there can be no doubt that they indicate the existence of heat in the interior of the earth. It may possibly be, as some have urged, that the volcanoes are merely vents for comparatively small masses of subterranean molten matter; it may be, as others more reasonably, in my opinion, believe, that the whole interior of the earth is at the temperature of incandescence, and that the eruptions of volcanoes and the shocks of earthquakes are merely consequences of the gradual shrinkage of the external crust, as it continually strives to accommodate [81] itself to the lessening bulk of the fluid interior. But whichever view we may adopt, it is at least obvious that the earth is in part, at all events, a heated body, and that the heat is not in the nature of a combustion, generated and sustained by the progress of chemical action. No doubt there may be local phenomena of this description, but by far the larger proportion of the earth's internal heat seems merely the fervour of incandescence. It is to be likened to the heat of the molten iron which has been run into the sand, rather than to the glowing coals in the furnace in which that iron has been smelted.

There is one volcanic outbreak of such exceptional interest in these modern times that I cannot refrain from alluding to it. Doubtless every one has heard of that marvellous eruption of Krakatoa, which occurred on August 26th and 27th, 1883, and gives a unique chapter in the history of volcanic phenomena. Not alone was the eruption of Krakatoa alarming in its more ordinary manifestations, but it was unparalleled both in the vehemence of the shock and in the distance to which the effects of the great eruption were propagated. [82] I speak not now of the great waves of ocean that inundated the coasts of Sumatra and Java, and swept away thirty-six thousand people, nor do I allude to the intense darkness which spread for one hundred and eighty miles or more all round. I shall just mention the three most important phenomena, which demonstrate the energy which still resides in the interior of our earth. Place a terrestrial globe before you, and fix your attention on the Straits of Sunda; think also of the great atmospheric ocean some two or three hundred miles deep which envelopes our earth. When a pebble is tossed into a pond a beautiful series of concentric ripples diverge from it; so when Krakatoa burst up in that mighty catastrophe, a series of gigantic waves were propagated through the air; they embraced the whole globe, converged to the antipodes of Krakatoa, thence again diverged, and returned to the seat of the volcano; a second time the mighty series of atmospheric ripples spread to the antipodes, and a second time returned. Seven times did that series of waves course over our globe, and leave their traces on every self-recording barometer that our earth possesses. [83] Thirty-six hours were occupied in the journey of the great undulation from Krakatoa to its antipodes. Perhaps even more striking was the extent of our earth's surface over which the noise of the great explosion spread.

At Batavia, ninety-four miles away, the concussions were simply deafening; at Macassar, in Celebes, two steamers were sent out to investigate the explosions which were heard, little thinking that they came from Krakatoa, nine hundred and sixty-nine miles away. Alarming sounds were heard over the island of Timor, one thousand three hundred and fifty-one miles away from Krakatoa. Diego Garcia in the Chogos islands is two thousand two hundred and sixty-seven miles from Krakatoa, but the thunders traversed even this distance, and were attributed to some ship in distress, for which a search was made. Most astounding of all, there is undoubted evidence that the sound of the mighty explosion was propagated across nearly the entire Indian ocean, and was heard in the island of Rodriguez, almost three thousand miles away. The immense distance over which this sound journeyed will be appreciated by the fact, that the noise did [84] not reach Rodriguez until four hours after it had left Krakatoa. In fact, it would seem that if Vesuvius were to explode with the same vehemence as Krakatoa did, the thunders of the explosion might penetrate so far as to be heard in London.

There is another and more beautiful manifestation of the world-wide significance of the Krakatoa outbreak. The vast column of smoke and ashes ascended twenty miles high in the air, and commenced a series of voyages around the equatorial regions of the earth. In three days it crossed the Indian ocean, and was traversing equatorial Africa; then came an Atlantic voyage; and then it coursed over central America, before a Pacific voyage brought it back to its point of departure after thirteen days; then the dust started again, and was traced around another similar circuit, while it was even tracked for a considerable time in placing the third girdle round the earth. Strange blue suns and green moons and other mysterious phenomena marked the progress of this vast volcanic cloud. At last the cloud began to lose its density, the dust spread more widely [85] over the tropics, became diffused through the temperate regions, and then the whole earth was able to participate in the glories of Krakatoa. The marvellous sunsets in the autumn of 1883 are attributable to this cause; and thus once again was brought before us the fact that the earth still contains large stores of thermal energy.

Attempts are sometimes made to explain volcanic phenomena on the supposition that they are entirely of a local character, and that we are not entitled to infer the incandescent nature of the earth's interior from the fact that volcanic outbreaks occasionally happen. For our present purpose this point is immaterial, though I must say it appears to me unreasonable to deny that the interior of the earth is in a most highly heated state. Every test we can apply shows us the existence of internal heat. Setting aside the more colossal phenomena of volcanic eruptions, we have innumerable minor manifestations of its presence. Are there not geysers and hot springs in many parts of the earth? and have we not all over our globe invariable testimony confirming the statement, that the deeper we go down beneath its [86] surface the hotter does the temperature become? Every miner is familiar with these facts; he knows that the deeper are his shafts the warmer it is down below, and the greater the necessity for providing increased ventilation to keep the temperature within a limit that shall be suitable for the workmen. All these varied classes of phenomena admit solely of one explanation, and that is, that the interior of the earth contains vast stores of incandescent heat.

We now apply to our earth the same reasoning which we should employ on a poker taken from the fire, or on a casting drawn from the foundry. Such bodies will lose their heat by radiation and conduction. The earth is therefore losing its heat. No doubt the process is an extremely slow one. The mighty reservoirs of internal heat are covered by vast layers of rock, which are such excellent non-conductors that they offer every possible impediment to the leakage of heat from the interior to the surface. We coat our steam-pipes over with non-conducting material, and this can now be done so successfully, that it is beginning to be found economical to transmit steam for a very long [87] distance through properly protected pipes. But no non-conducting material that we can manufacture can be half so effective as the shell of rock twenty miles or more in thickness, which secures the heated interior of the earth from rapid loss by radiation into space. Even were the earth's surface solid copper or solid silver, both most admirable conductors of heat, the cooling down of this vast globe would be an extremely tardy process; how much more tardy must it therefore be when such exceedingly bad conductors as rocks form the envelope? How imperfectly material of this kind will transmit heat is strikingly illustrated by the great blast iron furnaces which are so vitally important in one of England's greatest manufacturing industries. A glowing mass of coal and iron ore and limestone is here urged to vivid incandescence by a blast of air itself heated to an intense temperature. The mighty heat thus generated—sufficient as it is to detach the iron from its close alliance with the earthy materials and to render the metal out as a pure stream rushing white-hot from the vent—is sufficiently confined by a few feet of brick-work, one side of [88] which is therefore at the temperature of molten iron, while the other is at a temperature not much exceeding that of the air. We may liken the brick-work of a blast furnace to the rocky covering of the earth; in each case an exceedingly high temperature on one side is compatible with a very moderate temperature on the other.

Although the drainage of heat away from the earth's interior to its surface, and its loss there by radiation into space, is an extremely tardy process, yet it is incessantly going on. We have here again to note the ability for gigantic effect which a small but continually operating cause may have, provided it always tends in the same direction. The earth is incessantly losing heat; and though in a day, a week, or a year the loss may not be very significant, yet when we come to deal with periods of time that have to be reckoned by millions of years, it may well be that the effect of a small loss of heat per annum can, in the course of these ages, reach unimagined dimensions. Suppose, for instance, that the earth experienced a fall of temperature in its interior which amounted to only one-thousandth of a degree in a year. So minute a [89] quantity as this is imperceptible. Even in a century, the loss of heat at this rate would be only the tenth of a degree. There would be no possible way of detecting it; the most careful thermometer could not be relied on to tell us for a certainty that the temperature of the hot waters of Bath had declined the tenth of a degree; and I need hardly say, that the fall of a tenth of a degree would signify nothing in the lavas of Vesuvius, nor influence the thunders of Krakatoa by one appreciable note. So far as a human life or the life of the human race is concerned, the decline of a tenth of a degree per century in the earth's internal heat is absolutely void of significance. I cannot, however, impress upon you too strongly, that the mere few thousands of years with which human history is cognizant are an inappreciable moment in comparison with those unmeasured millions of years which geology opens out to us, or with those far more majestic periods which the astronomer demands for the events he has to describe.

An annual loss of even one-thousandth of a degree will be capable of stupendous achievements when supposed to operate during epochs of geological [90] magnitude. In fact, its effects would be so vast, that it seems hardly credible that the present loss of heat from the earth should be so great as to amount to an abatement of one-thousandth of a degree per annum, for that would mean, that in a thousand years the earth's temperature would decline by one degree, and in a million years the decline would amount to a thousand degrees. At all events, the illustration may suffice to show, that the fact that we are not able to prove by our instruments that the earth is cooling is no argument whatever against the inevitable law, that the earth, like every other heated body, must be tending towards a lower temperature.

Without pretending to any numerical accuracy, we can at all events give a qualitative if not a quantitative analysis of the past history of our earth, in so far as its changes of temperature are concerned. A million years ago our earth doubtless contained appreciably more heat than it does at present. I speak not now, of course, of mere solar heat — of the heat which gives us the vicissitudes of seasons; I am only referring to the original hoard of internal heat which is gradually waning. As therefore our [91] retrospect extends through millions and millions of past ages, we see our earth ever growing warmer and warmer the further and further we look back. There was a time when those heated strata which we have now to go deep down in mines to find were considerably nearer the surface. At present, were it not for the sun, the heat of the earth where we stand would hardly be appreciably above the temperature of infinite space — perhaps some 200 or 300 degrees below zero. But there must have been a time when there was sufficient internal heat to maintain the exterior at a warm and indeed at a very hot temperature. Nor is there any bound to our retrospect arising from the operation or intervention of any other agent, so far as we know; consequently the hotter and the hotter grows the surface the further and the further we look back. Nor can we stop until, at an antiquity so great that I do not venture on any estimate of the date, we discover that this earth must have consisted of glowing hot material. Further and further we can look back, and we see the rocks — or whatever other term we choose to apply to the then ingredients of the earth's crust — in a white-hot and even in a molten [92] condition. Thus our argument has led us to the belief that time was when this now solid globe of ours was a ball of white-hot fluid.

On the argument which I have here used there are just two remarks which I particularly wish to make. Note in the first place, that our reasoning is founded on the fact that the earth is at present, to some extent, heated. It matters not whether this heat be much or little; our argument would have been equally valid had the earth only contained a single particle of its mass at a somewhat higher temperature than the temperature of space. I am, of course, not alluding in this to heat which can be generated by combustion. The other point to which I refer is to remove an objection which may possibly be urged against this line of reasoning. I have argued that because the temperature is continually increasing as we look backwards, that therefore a very great temperature must once have prevailed. Without some explanation this argument is not logically complete. There is, it is well known, the old paradox of the geometric series; you may add a farthing to a halfpenny, and then a half-farthing, and then a [93] quarter-farthing, and then the eighth of a farthing, followed by the sixteenth, and thirty-second, and so on, halving the contribution each time. Now no matter how long you continue this process, even if you went on with it for ever, and thus made an infinite number of contributions, you would never accomplish the task of raising the original halfpenny to the dignity of a penny. An infinite number of quantities may therefore, as this illustration shows, never succeed in attaining any considerable dimensions. Our argument, however, with regard to the increase of heat as we look back is the very opposite of this. It is the essence of a cooling body to lose heat more rapidly in proportion as its temperature is greater. Thus though the one-thousandth of a degree may be all the fall of temperature that our earth now experiences in a twelvemonth, yet in those glowing days when the surface was heated to incandescence, the loss of heat per annum must have been immensely greater than it is now. It therefore follows that the rate of gain of the earth's heat as we look back must be of a different character to that of the geometric series which I have just illustrated; [94] for each addition to the earth's heat, as we look back from year to year, must grow greater and greater, and therefore there is here no shelter for a fallacy in the argument on which the existence of high temperature of primeval times is founded.

The reasoning that I have applied to our earth may be applied in almost similar words to the moon. It is true that we have not any knowledge of the internal nature of the moon at present, nor are we able to point to any active volcanic phenomena at present in progress there in support of the contention that the moon either has now internal heat, or did once possess it. It is, however, impossible to deny the evidence which the lunar craters afford as to the past existence of volcanic activity on our satellite. Heat, therefore, there was once in the moon; and accordingly we are enabled to conclude that, on a retrospect through illimitable periods of time, we must find the moon transformed from that cold and inert body she now seems to a glowing and incandescent mass of molten material. The earth therefore and the moon in some remote ages—not alone [95] anterior to the existence of life, but anterior even to the earliest periods of which geologists have cognizance—must have been both globes of molten materials which have consolidated into the rocks of the present epoch.

We must now revert to the tidal history of the earth-moon system. Did we not show that there was a time when the earth and the moon—or perhaps, I should say, the ingredients of the earth and moon—were close together, were indeed in actual contact? We have now learned, from a wholly different line of reasoning, that in very early ages both bodies were highly heated. Here as elsewhere in this theory we can make little or no attempt to give any chronology, or to harmonize the different lines along which the course of history has run. No one can form the slightest idea as to what the temperature of the earth and of the moon must have been in those primeval ages when they were in contact. It is impossible, however, to deny that they must both have been in a very highly heated state; and everything we know of the matter inclines us to the belief that the temperature of the earth-moon system must [96] at this critical epoch have been one of glowing incandescence and fusion. It is therefore quite possible that these bodies—the moon especially—may have then been not at all of the form we see them now. It has been supposed, and there are some grounds for the supposition, that at this initial stage of earth-moon history the moon materials did not form a globe, but were disposed in a ring which surrounded the earth, the ring being in a condition of rapid rotation. It was at a subsequent period, according to this view, that the substances in the ring gradually drew together, and then by their mutual attractions formed a globe which ultimately consolidated down into the compact moon as we now see it. I must, however, specially draw your attention to the clearly-marked line which divides the facts which dynamics have taught us from those notions which are to be regarded as more or less conjectural. Interpreting the action of the tides by the principles of dynamics, we are assured that the moon was once—or rather the materials of the moon—in the immediate vicinity of the earth. There, however, dynamics leaves us, and unfortunately withholds its accurate illumination [97] from the events which immediately preceded that state of things.

The theory of tidal evolution which I am describing in these lectures is mainly the work of Professor George H. Darwin of Cambridge. Much of the original parts of the theory of the tides was due to Sir William Thomson, and I have also mentioned how Professor Purser contributed an important element to the dynamical theory. It is, however, Darwin who has persistently deduced from the theory all the various consequences which can be legitimately drawn from it. Darwin, for instance, pointed out that as the moon is receding from us, it must, if we only look far enough back, have been once in practical contact with the earth. It is to Darwin also that we owe many of the other parts of a fascinating theory, either in its mathematical or astronomical aspect; but I must take this opportunity of saying, that I do not propose to make Professor Darwin or any of the other mathematicians I have named responsible for all that I shall say in these lectures. I must be myself accountable for the way in which the subject is being treated, as well as for many of [98] the illustrations used, and some of the deductions I have drawn from the subject.

It is almost unavoidable for us to make a surmise as to the cause by which the moon had come into this remarkable position close to the earth at the most critical epoch of earth-moon history.

With reference to this Professor Darwin has offered an explanation, which seems so exceedingly plausible that it is impossible to resist the notion that it must be correct. I will ask you to think of the earth not as a solid body covered largely with ocean, but as a glowing globe of molten material. In a globe of this kind it is possible for great undulations to be set up. Here is a large vase of water, and by displacing it I can cause the water to undulate with a period which depends on the size of the vessel; undulations can be set up in a bucket of water, the period of these undulations being dependent upon the dimensions of the bucket. Similarly in a vast globe of molten material certain undulations could be set up, and those undulations would have a period depending upon the dimensions of this vibrating mass. We may conjecture a mode in which such vibrations could be originated. [99] Imagine a thin shell of rigid material which just encases the globe; suppose this be divided into four quarters, like the four quarters of an orange, and that two of these opposite quarters be rejected, leaving two quarters on the liquid. Now suppose that these two quarters be suddenly pressed in, and then be as suddenly removed—they will produce depressions, of course, on the two opposite quarters, while the uncompressed quarters will become protuberant. In virtue of the mutual attractions between the different particles of the mass, an effort will be made to restore the globular form, but this will of course rather overshoot the mark; and therefore a series of undulations will be originated by which two opposite quarters of the sphere will alternately shrink in and become protuberant. There will be a particular period to this oscillation. For our globe it would appear to be somewhere about an hour and a half or two hours; but there is necessarily a good deal of uncertainty about this point.

We have seen how in those primitive days the earth was spinning around very rapidly; and I have also stated that the earth might at this very [100] critical epoch of its history be compared with a grindstone which is being driven so rapidly that it is on the very brink of rupture. It is remarkable to note, that a cause tending to precipitate a rupture of the earth was at hand. The sun then raised tides in the earth as it does at present. When the earth revolved in a period of some four hours or thereabouts, the high tides caused by the sun succeeded each other at intervals of about two hours. When I speak of tides in this respect, of course I am not alluding to oceanic tides; these were the days long before ocean existed, at least in the liquid form. The tides I am speaking about were raised in the fluids and materials which then constituted the whole of the glowing earth; those tides rose and fell under the throb produced by the sun, just as truly as tides produced in an ordinary ocean. But now note the significant coincidence between the period of the throb produced by the sun-raised tides, and that natural period of vibration which belonged to our earth as a mass of molten material. It therefore follows, that the impulse given to the earth by the sun harmonized in time with that period in which the earth itself [101] was disposed to oscillate. A well-known dynamical principle here comes into play. You see a heavy weight hanging by a string, and in my hand I hold a little slip of wood no heavier than a common pencil; ordinarily speaking, I might strike that heavy weight with this slip of wood, and no effect is produced; but if I take care to time the little blows that I give so that they shall harmonize with the vibrations which the weight is naturally disposed to make, then the effect of many small blows will be cumulative, so much so, that after a short time the weight begins to respond to my efforts, and now you see it has acquired a swing of very considerable amplitude. In Professor Fitzgerald's address to the British Association at Bath last autumn, he gives an account of those astounding experiments of Hertz, in which well-timed electrical impulses broke down an air resistance, and revealed to us ethereal vibrations which could never have been made manifest except by the principle we are here discussing. The ingenious conjecture has been made, that when the earth was thrown into tidal vibrations in those primeval days, these slight vibrations, harmonizing as they did [102] with the natural period of the earth, gradually acquired amplitude; the result being that the pulse of each successive vibration increased at last to such an extent that the earth separated under the stress, and threw off a portion of those semi-fluid materials of which it was composed. In process of time these rejected portions contracted together, and ultimately formed that moon we now see. Such is the origin of the moon which the modern theory of tidal evolution has presented to our notice.

There are two great epochs in the evolution of the earth-moon system — two critical epochs which possess a unique dynamical significance; one of these periods was early in the beginning, while the other cannot arrive for countless ages yet to come. I am aware that in discussing this matter I am entering somewhat largely into mathematical principles; I must only endeavour to state the matter as succinctly as the subject will admit.

In an earlier part of this lecture I have explained how, during all the development of the earth-moon system, the quantity of moment of momentum remains unaltered. The moment of [103] momentum of the earth's rotation added to the moment of momentum of the moon's revolution remains constant; if one of these quantities increase the other must decrease, and the progress of the evolution will have this result, that energy shall be gradually lost in consequence of the friction produced by the tides. The investigation is one appropriate for mathematical formulæ, such as those that can be found in Professor Darwin's memoirs; but nature has in this instance dealt kindly with us, for she has enabled an abstruse mathematical principle to be dealt with in a singularly clear and concise manner. We want to obtain a definite view of the alteration in the energy of the system which shall correspond to a small change in the velocity of the earth's rotation, the moon of course accommodating itself so that the moment of momentum shall be preserved unaltered. We can use for this purpose an angular velocity which represents the excess of the earth's rotation over the angular revolution of the moon; it is, in fact, the apparent angular velocity with which the moon appears to move round the heavens. If we represent by N the angular [104] velocity of the earth, and by M the angular velocity of the moon in its orbit round the earth, the quantity we desire to express is $N - M$; we shall call it the relative rotation. The mathematical theorem which tells us what we want can be enunciated in a concise manner as follows. *The alteration of the energy of the system may be expressed by multiplying the relative rotation by the change in the earth's angular velocity.* This result will explain many points to us in the theory, but just at present I am only going to make a single inference from it.

I must advert for a moment to the familiar conception of a maximum or a minimum. If a magnitude be increasing, that is, gradually growing greater and greater, it has obviously not attained a maximum so long as the growth is in progress. Nor if the object be actually decreasing can it be said to be at a maximum either; for then it was greater a second ago than it is now, and therefore it cannot be at a maximum at present. We may illustrate this by the familiar example of a stone thrown up into the air; at first it gradually rises, being higher at each instant than it was previously, [105] until a culminating point is reached, when just for a moment the stone is poised at the summit of its path ere it commences its return to earth again. In this case the maximum point is obtained when the stone, having ceased to ascend, and not having yet commenced to descend, is momentarily at rest.

The same principles apply to the determination of a minimum. As long as the magnitude is declining the minimum has not been reached; it is only when the decline has ceased, and an increase is on the point of setting in, that the minimum can be said to be touched.

The earth-moon system contains at any moment a certain store of energy, and to every conceivable condition of the earth-moon system a certain quantity of energy is appropriate. It is instructive for us to study the different positions in which the earth and the moon might lie, and to examine the different quantities of energy which the system will contain in each of those varied positions. It is however to be understood that the different cases all presuppose the same total moment of momentum.

Among the different cases that can be imagined, [106] those will be of special interest in which the total quantity of energy in the system is a maximum or a minimum. We must for this purpose suppose the system gradually to run through all conceivable changes, with the earth and moon as near as possible, and as far as possible, and in all intermediate positions; we must also attribute to the earth every variety in the velocity of its rotation which is compatible with the preservation of the moment of momentum. Beginning then with the earth's velocity of rotation at its lowest, we may suppose it gradually and continually increased, and, as we have already seen, the change in the energy of the system is to be expressed by multiplying the relative rotation into the change of the earth's angular velocity. It follows from the principles we have already explained, that the maximum or minimum energy is attained at the moment when the alteration is zero. It therefore follows, that the critical periods of the system will arise when the relative rotation is zero, that is, when the earth's rotation on its axis is performed with a velocity equal to that with which the moon revolves around the earth. This is truly a singular [107] condition of the earth-moon system; the moon in such a case would revolve around the earth as if the two bodies were bound together by rigid bonds into what was practically a single solid body. At the present moment no doubt to some extent this condition is realized, because the moon always turns the same face to the earth (a point on which we shall have something to say later on); but in the original condition of the earth-moon system, the earth would also constantly direct the same face to the moon, a condition of things which is now very far from being realized.

It can be shown from the mathematical nature of the problem that there are four states of the earth-moon system in which this condition may be realized, and which are also compatible with the conservation of the moment of momentum. We can express what this condition implies in a somewhat more simple manner. Let us understand by the *day* the period of the earth's rotation on its axis, whatever that may be, and let us understand by the *month* the period of revolution of the moon around the earth, whatever value it may have; then the condition of maximum or [108] minimum energy is attained when the day and the month have become equal to each other. Of the four occasions mathematically possible in which the day and the month can be equal, there are only two which at present need engage our attention—one of these occurred near the beginning of the earth and moon's history, the other remains to be approached in the immeasurably remote future. The two remaining solutions are futile, being what the mathematician would describe as imaginary.

There is a fundamental difference between the dynamical conditions in these critical epochs—in one of them the energy of the system has attained a maximum value, and in the other the energy of the system is at a minimum value. It is impossible to over-estimate the significance of these two states of the system.

I may recall the fundamental notion which every one has learned in mechanics, as to the difference between stable and unstable equilibrium. The conceivable possibility of making an egg stand on its end is a practical impossibility, because nature does not like unstable equilibrium, and a body [109] departs therefrom on the least disturbance; on the other hand, stable equilibrium is the position in which nature tends to place everything. A log of wood floating on a river might conceivably float in a vertical position with its end up out of the water, but you never could succeed in so balancing it, because no matter how carefully you adjusted the log, it would almost instantly turn over when you left it free; on the other hand, when the log floats naturally on the water it assumes a horizontal position, to which, when momentarily displaced therefrom, it will return if permitted to do so. We have here an illustration of the contrast between stable and unstable equilibrium. It will be found generally that a body is in equilibrium when its centre of gravity is at its highest point or at its lowest point; there is, however, this important difference, that when the centre of gravity is highest the equilibrium is unstable, and when the centre of gravity is lowest the equilibrium is stable. The potential energy of an egg poised on its end in unstable equilibrium is greater than when it lies on its side in stable equilibrium. In fact, energy must be [110] expended to raise the egg from the horizontal position to the vertical; while, on the other hand, work could conceivably be done by the egg when it passes from the vertical position to the horizontal. Speaking generally, we may say that the stable position indicates low energy, while a redundancy of that valuable agent is suggestive of instability.

We may apply similar principles to the consideration of the earth-moon system. It is true that we have here a series of dynamical phenomena, while the illustrations I have given of stable and unstable equilibrium relate only to statical problems; but we can have dynamical stability and dynamical instability, just as we can have stable and unstable equilibrium. Dynamical instability corresponds with the maximum of energy, and dynamical stability to the minimum of energy.

At that primitive epoch, when the energy of the earth-moon system was a maximum, the condition was one of dynamical instability; it was impossible that it should last. But now mark how truly critical an occurrence this must have been in the history of the earth-moon system, for have I not [111] already explained that it is a necessary condition of the progress of tidal evolution that the energy of the system should be always declining? But here our retrospect has conducted us back to a most eventful crisis, in which the energy was a maximum, and therefore cannot have been immediately preceded by a state in which the energy was greater still; it is therefore impossible for the tidal evolution to have produced this state of things; some other influence must have been in operation at this beginning of the earth-moon system.

Thus there can be hardly a doubt that immediately preceding the critical epoch the moon originated from the earth in the way we have described. Note also that this condition, being one of maximum energy, was necessarily of dynamical instability, it could not last; the moon must adopt either of two courses—it must tumble back on the earth, or it must start outwards. Now which course was the moon to adopt? The case is analogous to that of an egg standing on its end—it will inevitably tumble one way or the other. Some infinitesimal cause will produce [112] a tendency towards one side, and to that side accordingly the egg will fall. The earth-moon system was similarly in an unstable state, an infinitesimal cause might conceivably decide the fate of the system. We are necessarily in ignorance of what the determining cause might have been, but the effect it produced is perfectly clear; the moon did not again return to its mother earth, but set out on that mighty career which is in progress to-day.

Let it be noted that these critical epochs in the earth-moon history arise when and only when there is an absolute identity between the length of the month and the length of the day. It may be proper therefore that I should provide a demonstration of the fact, that the identity between these two periods must necessarily have occurred at a very early period in the evolution.

The law of Kepler, which asserts that the square of the periodic time is proportioned to the cube of the mean distance, is in its ordinary application confined to a comparison between the revolutions of the several planets about the sun. The periodic [113] time of each planet is connected with its average distance by this law; but there is another application of Kepler's law which gives us information of the distance and the period of the moon in former stages of the earth-moon history. Although the actual path of the moon is of course an ellipse, yet that ellipse is troubled, as is well known, by many disturbing forces, and from this cause alone the actual path of the moon is far from being any of those simple curves with which we are so well acquainted. Even were the earth and the moon absolutely rigid particles, perturbations would work all sorts of small changes in the pliant curve. The phenomena of tidal evolution impart an additional element of complexity into the actual shape of the moon's path. We now see that the ellipse is not merely subject to incessant deflections of a periodic nature, it also undergoes a gradual contraction as we look back through time past; but we may, with all needful accuracy for our present purpose, think of the path of the moon as a circle, only we must attribute to that circle a continuous contraction of its radius the further and the further we look back. The alteration [114] in the radius will be even so slow, that the moon will accomplish thousands of revolutions around the earth without any appreciable alteration in the average distance of the two bodies. We can therefore think of the moon as revolving at every epoch in a circle of special radius, and as accomplishing that revolution in a special time. With this understanding we can now apply Kepler's law to the several stages of the moon's past history. The periodic time of each revolution, and the mean distance at which that revolution was performed, will be always connected together by the formula of Kepler. Thus to take an instance in the very remote past. Let us suppose that the moon was at one hundred and twenty thousand miles instead of two hundred and forty thousand, that is, at half its present distance. Applying the law of Kepler, we see that the time of revolution must then have been only about ten days instead of the twenty-seven it is now. Still further, let us suppose that the moon revolves in an orbit with one-tenth of the diameter it has at present, then the cube of 10 being 1000, and the square root of 1000 being 31.6, it follows that the month must have [115] been less than the thirty-first part of what it is at present, that is, it must have been considerably less than one of our present days. Thus you see the month is growing shorter and shorter the further we look back, the day is also growing shorter and shorter; but still I think we can show that there must have been a time when the month will have been at least as short as the day. For let us take the most extreme case in which the moon shall have made the closest possible approximation to the earth. Two globes in contact will have a distance between their centres which is equal to the sum of their radii. Take the earth as having a radius of four thousand miles, and the moon a radius of one thousand miles, the two centres must at their shortest distance be five thousand miles apart, that is, the moon must then be at the forty-eighth part of its present distance from the earth. Now the cube of 48 is 110,592, and the square root of 110,592 is nearly 333, therefore the length of the month will be one-three hundred and thirty-third part of the duration of the month at present; in other words, the moon must revolve around the earth in a period of somewhat about two hours. It [116] seems impossible that the day can ever have been as brief as this. We have therefore proved that, in the course of its contracting duration, the moon must have overtaken the contracting day, and that therefore there must have been a time when the moon was in the vicinity of the earth, and having a day and month of equal period. Thus we have shown that the critical condition of dynamical instability must have occurred in the early period of the earth-moon history, if the agents then in operation were those which we now know. The further development of the subject must be postponed until the next lecture.

LECTURE II. [117]

Starting from that fitting commencement of earth-moon history which the critical epoch affords, we shall now describe the dynamical phenomena as the tidal evolution progressed. The moon and the earth initially moved as a solid body, each bending the same face towards the other; but as the moon retreated, and as tides began to be raised on the earth, the length of the day began to increase, as did also the length of the month. We know, however, that the month increased more rapidly than the day, so that a time was reached when the month was twice as long as the day; and still both periods kept on increasing, but not at equal rates, for in progress of time the month grew so much more rapidly than the day, that many days had to elapse while the moon accomplished a single revolution. It is, however, only necessary for us to note those stages of the mighty [118] progress which correspond to special events. The first of such stages was attained when the month assumed its maximum ratio to the day. At this time, the month was about twenty-nine days, and the epoch appears to have occurred at a comparatively recent date if we use such standards of time as tidal evolution requires; though measured by historical standards, the epoch is of incalculable antiquity. I cannot impress upon you too often the enormous magnitude of the period of time which these phenomena have required for their evolution. Professor Darwin's theory affords but little information on this point, and the utmost we can do is to assign a minor limit to the period through which tidal evolution has been in progress. It is certain that the birth of the moon must have occurred at least fifty million years ago, but probably the true period is enormously greater than this. If indeed we choose to add a cipher or two to the figure just printed, I do not think there is anything which could tell us that we have over-estimated the mark. Therefore, when I speak of the epoch in which the month possessed the greatest number of [119] days as a recent one, it must be understood that I am merely speaking of events in relation to the order of tidal evolution. Viewed from this standpoint, we can show that the epoch is a recent one in the following manner. At present the month consists of a little more than twenty-seven days, but at this maximum period to which I have referred the month was about twenty-nine days; from that it began to decline, and the decline cannot have proceeded very far, for even still there are only two days less in the month than at the time when the month had the greatest number of days. It thus follows that the present epoch—the human epoch, as we may call it—in the history of the earth has fallen at a time when the progress of tidal evolution is about half-way between the initial and the final stage. I do not mean half-way in the sense of actual measurement of years; indeed, from this point it would seem that we cannot yet be nearly half-way, for, vast as are the periods of time that have elapsed since the moon first took its departure from the earth, they fall far short of that awful period of time which will intervene between the present moment and the hour [120] when the next critical state of earth-moon history shall have been attained. In that state the day is destined once again to be equal to the month, just as was the case in the initial stage. The half-way stage will therefore in one sense be that in which the proportion of the month to the day culminates. This is the stage which we have but lately passed; and thus it is that at present we may be said to be almost half-way through the progress of tidal evolution.

My narrative of the earth-moon evolution must from this point forward cease to be retrospective. Having begun at that critical moment when the month and day were first equal, we have traced the progress of events to the present hour. What we have now to say is therefore a forecast of events yet to come. So far as we can tell, no agent is likely to interfere with the gradual evolution caused by the tides, which dynamical principles have disclosed to us. As the years roll on, or perhaps, I should rather say, as thousands of years and millions of years roll on, the day will continue to elongate, or the earth to rotate more slowly on its axis. But countless ages must [121] elapse before another critical stage of the history shall be reached. It is needless for me to ponder over the tedious process by which this interesting epoch is reached. I shall rather sketch what the actual condition of our system will be when that moment shall have arrived. The day will then have expanded from the present familiar twenty-four hours up to a day more than twice, more than five, even more than fifty times its present duration. In round numbers, we may say that this great day will occupy one thousand four hundred of our ordinary hours. To realize the critical nature of the situation then arrived at, we must follow the corresponding evolution through which the moon passes. From its present distance of two hundred and forty thousand miles, the moon will describe an ever-enlarging orbit; and as it does so the duration of the month will also increase, until at last a point will be reached when the month has become more than double its present length, and has attained the particular value of one thousand four hundred hours. We are specially to observe that this one-thousand four-hundred-hour month will be exactly reached [122] when the day has also expanded to one thousand four hundred hours; and the essence of this critical condition, which may be regarded as a significant point of tidal evolution, is that the day and the month have again become equal. The day and the month were equal at the beginning, the day and the month will be equal at the end. Yet how wide is the difference between the beginning and the end. The day or the month at the end is some hundreds of times as long as the month or the day at the beginning.

I have already fully explained how, in any stage of the evolutionary progress in which the day and the month became equal, the energy of the system attained a maximum or a minimum value. At the beginning the energy was a maximum; at the end the energy will be a minimum. The most important consequences follow from this consideration. I have already shown that a condition of maximum energy corresponded to dynamic instability. Thus we saw that the earth-moon history could not have commenced without the intervention of some influence other than tides at the beginning. Now let us learn what the similar [123] doctrine has to tell us with regard to the end. The condition then arrived at is one of dynamical stability; for suppose that the system were to receive a slight alteration, by which the moon went out a little further, and thus described a larger orbit, and so performed more than its share of the moment of spin. Then the earth would have to do a little less spinning, because, under all circumstances, the total quantity of spin must be preserved unaltered. But the energy being at a minimum, such a small displacement must of course produce a state of things in which the energy would be increased. Or if we conceived the moon to come in towards the earth, the moon would then contribute less to the total moment of momentum. It would therefore be incumbent on the earth to do more; and accordingly the velocity of the earth's rotation would be augmented. But this arrangement also could only be produced by the addition of some fresh energy to the system, because the position from which the system is supposed to have been disturbed is one of minimum energy.

No disturbance of the system from this final [124] position is therefore conceivable, unless some energy can be communicated to it. But this will demonstrate the utter incompetency of the tides to shift the system by a hair's breadth from this position; for it is of the essence of the tides to waste energy by friction. And the transformations of the system which the tides have caused are invariably characterized by a decline of energy, the movements being otherwise arranged so that the total moment of momentum shall be preserved intact. Note, how far we were justified in speaking of this condition as a final one. It is final so far as the lunar tides are concerned; and were the system to be screened from all outer interference, this accommodation between the earth and the moon would be eternal.

There is indeed another way of demonstrating that a condition of the system in which the day has assumed equality with the month must necessarily be one of dynamical equilibrium. We have shown that the energy which the tides demand is derived not from the mere fact that there are high tides and low tides, but from the circumstance that these tides do rise and fall; that in falling and [125] rising they do produce currents; and it is these currents which generate the friction by which the earth's velocity is slowly abated, its energy wasted, and no doubt ultimately dissipated as heat. If therefore we can make the ebbing and the flowing of the tides to cease, then our argument will disappear. Thus suppose, for the sake of illustration, that at a moment when the tides happened to be at high water in the Thames, such a change took place in the behaviour of the moon that the water always remained full in the Thames, and at every other spot on the earth remained fixed at the exact height which it possessed at this particular moment. There would be no more tidal friction, and therefore the system would cease to course through that series of changes which the existence of tidal friction necessitates.

But if the tide is always to be full in the Thames, then the moon must be always in the same position with respect to the meridian, that is, the moon must always be fixed in the heavens over London. In fact, the moon must then revolve around the earth just as fast as London does—the month must have the same length as the day. The earth must then [126] show the same face constantly to the moon, just as the moon always does show the same face towards the earth; the two globes will in fact revolve as if they were connected with invisible bonds, which united them into a single rigid body.

We need therefore feel no surprise at the cessation of the progress of tidal evolution when the month and the day are equal, for then the movement of moon-raised tides has ceased. No doubt the same may be said of the state at the beginning of the history, when the day and the month had the brief and equal duration of a few hours. While the equality of the two periods lasted there could be no tides, and therefore no progress in the direction of tidal evolution. There is, however, the profound difference of stability and instability between the two cases; the most insignificant disturbance of the system at the initial stage was sufficient to precipitate the revolving moon from its condition of dynamical equilibrium, and to start the course of tidal evolution in full vigour. If, however, any trifling derangement should take place in the last condition of the system, so that the month and the day departed slightly from [127] equality, there would instantly be an ebbing and a flowing of the tides; and the friction generated by these tides would operate to restore the equality because this condition is one of dynamical stability.

It will thus be seen with what justice we can look forward to the day and month each of fourteen hundred hours as a finale to the progress of the luni-tidal evolution. Throughout the whole of this marvellous series of changes it is always necessary to remember the one constant and invariable element—the moment of momentum of the system which tides cannot alter. Whatever else the friction can have done, however fearful may have been the loss of energy by the system, the moment of momentum which the system had at the beginning it preserves unto the end. This it is which chiefly gives us the numerical data on which we have to rely for the quantitative features of tidal evolution.

We have made so many demands in the course of these lectures on the capacity of tidal friction to accomplish startling phenomena in the evolution of the earth-moon system, that it is well for us to seek for any evidence that may otherwise be obtainable as to the capacity of tides for the [128] accomplishment of gigantic operations. I do not say that there is any doubt which requires to be dispelled by such evidence, for as to the general outlines of the doctrine of tidal evolution which has been here sketched out there can be no reasonable ground for mistrust; but nevertheless it is always desirable to widen our comprehension of any natural phenomena by observing collateral facts. Now there is one branch of tidal action to which I have as yet only in the most incidental way referred. We have been speaking of the tides in the earth which are made to ebb and flow by the action of the moon; we have now to consider the tides in the moon, which are there excited by the action of the earth. For between these two bodies there is a reciprocity of tidal-making energy—each of them is competent to raise tides in the other. As the moon is so small in comparison with the earth, and as the tides on the moon are of but little significance in the progress of tidal evolution, it has been permissible for us to omit them from our former discussion. But it is these tides on the moon which will afford us a striking illustration of the competency of tides for stupendous [129] tasks. The moon presents a monument to show what tides are able to accomplish.

Fig. 3.—The Moon.

I must first, however, explain a difficulty which is almost sure to suggest itself when we speak of tides on the moon. I shall be told that the moon contains no water on its surface, and how then, it will be said, can tides ebb and flow where there is no sea to be disturbed? There are two answers to this difficulty; it is no doubt true that the moon seems at present entirely devoid of water in so far as its surface is exposed to us, but it is by no means certain that the moon was always in this destitute condition. There are very large features marked on its map as "seas"; these regions are of a darker hue than the rest of the moon's surface, they are large objects often many hundreds of miles in diameter, and they form, in fact, those dark patches on the brilliant surface which are conspicuous to the unaided eye, and are represented in Fig. 3. Viewed in a telescope these so-called seas, while clearly possessing no water at the present time, are yet widely different from the general aspect of the moon's surface. It has often been supposed that great oceans once filled [130] these basins, and a plausible explanation has even been offered as to how the waters they once contained could have vanished. It has been thought that as the mineral substances deep in [131] the interior of our satellite assumed the crystalline form during the progress of cooling, the demand for water of crystallization required for incorporation with the minerals was so great that the oceans of the moon became entirely absorbed. It is, however, unnecessary for our present argument that this theory should be correct. Even if there never was a drop of water found on our satellite, the tides in its molten materials would be quite sufficient for our purpose; anything that tides could accomplish would be done more speedily by vast tides of flowing lava than by merely oceanic tides.

There can be no doubt that tides raised on the moon by the earth would be greater than the tides raised on the earth by the moon. The question is, however, not a very simple one, for it depends on the masses of both bodies as well as on their relative dimensions. In so far as the masses are concerned, the earth being more than eighty times as heavy as the moon, the tides would on this account be vastly larger on the moon than on the earth. On the other hand, the moon's diameter being much less than that of the earth, the efficiency [132] of a tide-producing body in its action on the moon would be less than that of the same body at the same distance in its action on the earth; but the diminution of the tides from this cause would be not so great as their increase from the former cause, and therefore the net result would be to exhibit much greater tides on the moon than on the earth.

Suppose that the moon had been originally endowed with a rapid movement of rotation around its axis, the effect of the tides on that rotation would tend to check its velocity just in the same way as the tides on the earth have effected a continual elongation of the day. Only as the tides on the moon were so enormously great, their capacity to check the moon's speed would have corresponding efficacy. In addition to this, the mass of the moon being so small, it could only offer feeble resistance to the unceasing action of the tide, and therefore our satellite must succumb to whatever the tides desired ages before our earth would have been affected to a like extent. It must be noticed that the influence of the tidal friction is not directed to the total annihilation of [133] the rotation of the two bodies affected by it; the velocity is only checked down until it attains such a point that the speed in which each body rotates upon its axis has become equal to that in which it revolves around the tide-producer. The practical effect of such an adjustment is to make the tide-agitated body turn a constant face towards its tormentor.

I may here note a point about which people sometimes find a little difficulty. The moon constantly turns the same face towards the earth, and therefore people are sometimes apt to think that the moon performs no rotation whatever around its own axis. But this is indeed not the case. The true inference to be drawn from the constant face of the moon is, that the velocity of rotation about its own axis is equal to that of its rotation around the earth; in fact, the moon revolves around the earth in twenty-seven days, and its rotation about its axis is performed in twenty-seven days also. You may illustrate the movement of the moon around the earth by walking around a table in a room, keeping all the time your face turned towards the table; in such a case [134] as this you not only perform a motion of revolution, but you also perform a rotation in an equal period. The proof that you do rotate is to be found in the fact that during the movement your face is being directed successively to all the points of the compass. There is no more singular fact in the solar system than the constancy of the moon's face to the earth. The periods of rotation and revolution are both alike; if one of these periods exceeded the other by an amount so small as the hundredth part of a second, the moon would in the lapse of ages permit us to see that other side which is now so jealously concealed. The marvellous coincidence between these two periods would be absolutely inexplicable, unless we were able to assign it to some physical cause. It must be remembered that in this matter the moon occupies a unique position among the heavenly host. The sun revolves around on its axis in a period of twenty-five or twenty-six days — thus we see one side of the sun as frequently as we see the other. The side of the sun which is turned towards us to-day is almost entirely different from that we saw a fortnight ago. Nor is the period of the sun's rotation to be identified [135] with any other remarkable period in our system. If it were equal to the length of the year, for instance, or if it were equal to the period of any of the other planets, then it could hardly be contended that the phenomenon as presented by the moon was unique; but the sun's period is not simply related, or indeed related at all, to any of the other periodic times in the system. Nor do we find anything like the moon's constancy of face in the behaviour of the other planets. Jupiter turns now one face to us and then another. Nor is his rotation related to the sun or related to any other body, as our moon's motion is related to us. It has indeed been thought that in the movements of the satellites of Jupiter a somewhat similar phenomenon may be observed to that in the motion of our own satellite. If this be so, the causes whereby this phenomenon is produced are doubtless identical in the two cases.

So remarkable a coincidence as that which the moon's motion shows could not reasonably be explained as a mere fortuitous circumstance; nor need we hesitate to admit that a physical explanation is required when we find a most satisfactory [136] one ready for our acceptance, as was originally pointed out by Helmholtz.

There can be no doubt whatever that the constancy of the moon's face is the work of ancient tides, which have long since ceased to act. We have shown that if the moon's rotation had once been too rapid to permit of the same face being always directed towards us, the tides would operate as a check by which the velocity of that rotation would be abated. On the other hand, if the moon rotated so slowly that its other face would be exposed to us in the course of the revolution, the tides would then be dragged violently over its surface in the direction of its rotation; their tendency would thus be to accelerate the speed until the angular velocity of rotation was equal to that of revolution. Thus the tides would act as a controlling agent of the utmost stringency to hurry the moon round when it was not turning fast enough, and to arrest the motion when going too fast. Peace there would be none for the moon until it yielded absolute compliance to the tyranny of the tides, and adjusted its period of rotation with exact identity to its period of revolution. Doubtless this adjustment [137] was made countless ages ago, and since that period the tides have acted so as to preserve the adjustment, as long as any part of the moon was in a state sufficiently soft or fluid to respond to tidal impression. The present state of the moon is a monument to which we may confidently appeal in support of our contention as to the great power of the tides during the ages which have passed; it will serve as an illustration of the future which is reserved for our earth in ages yet to come, when our globe shall have also succumbed to tidal influence.

It is owing to the smallness of the moon relatively to the earth that the tidal process has reached a much more advanced stage in the moon than it has on the earth; but the moon is incessant in its efforts to bring the earth into the same condition which it has itself been forced to assume. Thus again we look forward to an epoch in the inconceivably remote future when tidal thraldom shall be supreme, and when the earth shall turn the same face to the moon, as the moon now turns the same face to the earth.

In the critical state of things thus looming in [138] the dim future, the earth and the moon will continue to perform this adjusted revolution in a period of about fourteen hundred hours, the two bodies being held, as it were, by invisible bands. Such an arrangement might be eternal if there were no intrusion of tidal influence from any other body; but of course in our system as we actually find it the sun produces tides as well as the moon; and the solar tides being at present much less than those originated by the moon, we have neglected them in the general outlines of the theory. The solar tides, however, must necessarily have an increasing significance. I do not mean that they will intrinsically increase, for there seems no reason to apprehend any growth in their actual amount; it is their relative importance to the lunar tides that is the augmenting quantity. As the final state is being approached, and as the velocity of the earth's rotation is approximating to the angular velocity with which the moon revolves around it, the ebbing and the flowing of the lunar tides must become of evanescent importance; and this indeed for a double reason, partly on account of the moon's greatly augmented distance, and [139] partly on account of the increasing length of the lunar day, and the extremely tardy movements of ebb and flow that the lunar tides will then have. Thus the lunar tides, so far as their dynamical importance is concerned, will ultimately become zero, while the solar tides retain all their pristine efficiency.

We have therefore to examine the dynamical effects of solar tides on the earth and moon in the critical stage to which the present course of things tends. The earth will then rotate in a period of about fifty-seven of its present days; and considering that the length of the day, though so much greater than our present day, is still much less than the year, it follows that the solar tides must still continue so as to bring the earth's velocity of rotation to a point even lower than it has yet attained. In fact, if we could venture to project our glance sufficiently far into the future, it would seem that the earth must ultimately have its velocity checked by the sun-raised tides, until the day itself had become equal to the year. The dynamical considerations become, however, too complex for us to follow them, so [140] that I shall be content with merely pointing out that the influence of the solar tides will prevent the earth and moon from eternally preserving the relations of bending the same face towards each other; the earth's motion will, in fact, be so far checked, that the day will become *longer* than the month.

Thus the doctrine of tidal evolution has conducted us to a prospect of a condition of things which will some time be reached, when the moon will have receded to a distance in which the month shall have become about fifty-seven days, and when the earth around which this moon revolves shall actually require a still longer period to accomplish its rotation on its axis. Here is an odd condition for a planet with its satellite; indeed, until a dozen years ago it would have been pronounced inconceivable that a moon should whirl round a planet so quickly that its journey was accomplished in less than one of the planet's own days. Arguments might be found to show that this was impossible, or at least unprecedented. There is our own moon, which now takes twenty-seven days to go round the earth; there is Jupiter, with four moons, and the nearest of these to the [141] primary goes round in forty-two and a half hours. No doubt this is a very rapid motion; but all those matters are much more lively with Jupiter than they are here. The giant planet himself does not need ten hours for a single rotation, so that you see his nearest moon still takes between two and three Jovian days to accomplish a single revolution. The example of Saturn might have been cited to show that the quickest revolution that any satellite could perform must still require at least twice as long as the day in which the planet performed its rotation. Nor could the rotation of the planets around the sun afford a case which could be cited. For even Mercury, the nearest of all the planets to the sun of which the existence is certainly known, and therefore the most rapid in its revolution, requires eighty-eight days to get round once; and in the mean time the sun has had time to accomplish between three and four rotations. Indeed, the analogies would seem to have shown so great an improbability in the conclusion towards which tidal evolution points, that they would have contributed a serious obstacle to the general acceptance of that theory. [142]

But in 1877 an event took place so interesting in astronomical history, that we have to look back to the memorable discovery of Uranus in 1781 before we can find a parallel to it in importance. Mars had always been looked upon as one of the moonless planets, though grounds were not wanting for the surmise that probably moons to Mars really existed. It was under the influence of this belief that an attempt was made by Professor Asaph Hall at Washington to make a determined search, and see if Mars might not be attended by satellites large enough to be discoverable. The circumstances under which this memorable inquiry was undertaken were eminently favourable for its success. The orbit of Mars is one which possesses an exceptionally high eccentricity; it consequently happens that the oppositions during which the planet is to be observed vary very greatly in the facilities they afford for a search like that contemplated by Professor Hall. It is obviously advantageous that the planet should be situated as near as possible to the earth, and in the opposition in 1877 the distance was almost at the lowest point it is capable of attaining; but this [143] was not the only point in which Professor Hall was favoured; he had the use of a telescope of magnificent proportions and of consummate optical perfection. His observatory was also placed in Washington, so that he had the advantage of a pure sky and of a much lower latitude than any observatory in Great Britain is placed at. But the most conspicuous advantage of all was the practised skill of the astronomer himself, without which all these other advantages would have been but of little avail. Great success rewarded his well-designed efforts; not alone was one satellite discovered which revolved around the planet in a period conformable with that of other similar cases, but a second little satellite was found, which accomplished its revolution in a wholly unexpected and unprecedented manner. The day of

Mars himself, that is, the period in which he can accomplish a rotation around his axis, very closely approximates to our own day, being in fact half an hour longer. This little satellite, the inner and more rapid of the pair, requires for a single revolution a period of only seven hours thirty-nine minutes, that is to say, the little body scampers more than three [144] times round its primary before the primary itself has finished one of its leisurely rotations. Here was indeed a striking fact, a unique fact in our system, which riveted the attention of astronomers on this most beautiful discovery.

You will now see the bearing which the movement of the inner satellite of Mars has on the doctrine of tidal evolution. As a legitimate consequence of that doctrine, we came to the conclusion that our earth-moon system must ultimately attain a condition in which the day is longer than the month. But this conclusion stood unsupported by any analogous facts in the more anciently-known truths of astronomy. The movement of the satellite of Mars, however, affords the precise illustration we want; and this fact, I think, adds an additional significance to the interest and the beauty of Professor Hall's discovery.

It is of particular interest to investigate the possible connection which the phenomena of tidal evolution may have had in connection with the geological phenomena of the earth. We have already pointed out the greater closeness of the moon to us in times past. The tides raised by the [145] moon on the earth must therefore have been greater in past ages than they are now, for of course the nearer the moon the bigger the tide. As soon as the earth and the moon had separated to a considerable distance we may say that the height of the tide will vary inversely as the cube of the moon's distance; it will therefore happen, that when the moon was at half its present distance from us, his tide-producing capacity was not alone twice as much or four times as much, but even eight times as much as it is at present; and a much greater rate of tidal rise and fall indicates, of course, a preponderance in every other manifestation of tidal activity. The tidal currents, for instance, must have been much greater in volume and in speed; even now there are places in which the tidal currents flow at four or more miles per hour. We can imagine, therefore, the vehemence of the tidal currents which must have flowed in those days when the moon was a much smaller distance from us. It is interesting to view these considerations in their possible bearings on geological phenomena. It is true that we have here many elements of uncertainty, but there is, however, [146] a certain general outline of facts which may be laid down, and which appears to be instructive, with reference to the past history of our earth.

I have all through these lectures indicated a mighty system of chronology for the earth-moon system. It is true that we cannot give our chronology any accurate expression in years. The various stages of this history are to be represented by the successive distances between the earth and the moon. Each successive epoch, for instance, may be marked by the number of thousands of miles which separate the moon from the earth.

But we have another system of chronology derived from a wholly different system of ideas; it too relates to periods of vast duration, and, like our great tidal periods, extends to times anterior to human history, or even to the duration of human life on this globe. The facts of geology open up to us a majestic chronology, the epochs of which are familiar to us by the succession of strata forming the crust of the earth, and by the succession of living beings whose remains these strata have preserved. From the present or recent age our retrospect over geological chronology leads [147] us to look through a vista embracing periods of time overwhelming in their duration, until at last our view becomes lost, and our imagination is baffled in the effort to comprehend the formation of those vast stratified rocks, a dozen miles or more in thickness, which seem to lie at the very base of the stratified system on the earth, and in which it would appear that the dawnings of life on this globe may be almost discerned. We have thus the two systems of chronology to compare — one, the astronomical chronology measured by the successive stages in the gradual retreat of the moon; the other, the geological chronology measured by the successive strata constituting the earth's crust. Never was a more noble problem proposed in the physical history of our earth than that which is implied in the attempt to correlate these two systems of chronology. What we would especially desire to know is the moon's distance which corresponds to each of the successive strata on the earth. How far off, for instance, was that moon which looked down on the coal forests in the time of their greatest luxuriance? or what was the apparent size of the full moon at which [148] the ichthyosaurus could have peeped when he turned that wonderful eye of his to the sky on a fine evening? But interesting as this great problem is, it lies, alas! outside the possibility of exact solution. Indeed we shall not make any attempt which must necessarily be futile to correlate these chronologies; all we can do is to state the one fact which is absolutely undeniable in the matter.

Let us fix our attention on that specially interesting epoch at the dawn of geological time, when those mighty Laurentian rocks were deposited of which the thickness is so astounding, and let us consider what the distance of the moon must have been at this initial epoch of the earth's history. All we know for certain is, that the moon must have been nearer, but what proportion that distance bore to the present distance is necessarily quite uncertain. Some years ago I delivered a lecture at Birmingham, entitled "A Glimpse through the Corridors of Time," and in that lecture I threw out the suggestion that the moon at this primeval epoch may have only been at a small fraction of its present distance from us, and that consequently [149] terrific tides may in these days have ravaged the coast. There was a good deal of discussion on the subject, and while it was universally admitted that the tides must have been larger in palæozoic times than they are at present, yet there was a considerable body of opinion to the effect that the tides even then may have been only about twice, or possibly not so much, greater than those tides we have at the present. What the actual fact may be we have no way of knowing; but it is interesting to note that even the smallest accession to the tides would be a valuable factor in the performance of geological work.

For let me recall to your minds a few of the fundamental phenomena of geology. Those stratified rocks with which we are now concerned have been chiefly manufactured by deposition of sediment in the ocean. Rivers, swollen, it may be, by floods, and turbid with a quantity of material held in suspension, discharge their waters into the sea. Granting time and quiet, this sediment falls to the bottom; successive additions are made to its thickness during centuries and thousands of years, and thus beds are formed which in the course of [150] ages consolidate into actual rock. In the formation of such beds the tides will play a part. Into the estuaries at the mouths of rivers the tides hurry in and hurry out, and especially during spring tides there are currents which flow with tremendous power; then too, as the waves batter against the coast they gradually wear away and crumble down the mightiest cliffs, and waft the sand and mud thus produced to augment that which has been brought down by the rivers. In this operation also the tides play a part of conspicuous importance, and where the ebb and flow is greatest it is obvious that an additional impetus will be given to the manufacture of stratified rocks. In fact, we may regard the waters of the globe as a mighty mill, incessantly occupied in grinding up materials for future strata. The main operating power of this mill is of course derived from the sun, for it is the sun which brings up the rains to nourish the rivers, it is the sun which raises the wind which lashes the waves against the shore. But there is an auxiliary power to keep the mill in motion, and that auxiliary power is afforded by the tides. If then we find that by any cause the efficiency [151] of the tides is increased we shall find that the mill for the manufacture of strata obtains a corresponding accession to its capacity. Assuming the estimate of Professor Darwin, that the tide may have had twice as great a vertical range of ebb and flow within geological times as it has at present, we find a considerable addition to the efficiency of the ocean in the manufacture of the ancient stratified rocks. It must be remembered that the extent of the area through which the tides will submerge and lay bare the country, will often be increased more than twofold by a twofold increase of height. A little illustration may show what I mean. Suppose a cone to be filled with water up to a certain height, and that the quantity of water in it be measured; now let the cone be filled until the water is at double the depth; then the surfaces of the water in the two cases will be in the ratio of the circles, one of which has double the diameter of the other. The areas of the two surfaces are thus as four to one; the volumes of the waters in the two cases will be in the proportion of two similar solids, the ratios of their dimensions being as two to one. Of course this means that [152] the water in the one case would be eight times as much as in the other. This particular illustration will not often apply exactly to tidal phenomena, but I may mention one place that I happen to know of, in the vicinity of Dublin, in which the effect of the rise and fall of the tide would be somewhat of this description. At Malahide there is a wide shallow estuary cut off from the sea by a railway embankment, and there is a viaduct in the embankment through which a great tidal current flows in and out alternately. At low tide there is but little water in this estuary, but at high tide it extends for miles inland. We may regard this inlet with sufficient approximation to the truth as half of a cone with a very large angle, the railway embankment of course forming the diameter; hence it follows that if the tide was to be raised to double its height, so large an area of additional land would be submerged, and so vast an increase of water would be necessary for the purpose, that the flow under the railway bridge would have to be much more considerable than it is at present. In some degree the same phenomena will be repeated elsewhere around the coast. Simply [153] multiplying the height of the tide by two would often mean that the border of land between high and low water would be increased more than twofold, and that the volume of water alternately poured on the land and drawn off it would be increased in a still larger proportion. The velocity of all tidal currents would also be greater than at present, and as the power of a current of water for transporting solid material held in suspension increases rapidly with the velocity, so we may infer that the efficiency of tidal currents as a vehicle for the transport of comminuted rocks would be greatly increased. It is thus obvious that tides with a rise and fall double in vertical height of those which we know at present would add a large increase to their efficiency as geological agents. Indeed, even were the tides only half or one-third greater than those we know now, we might reasonably expect that the manufacture of stratified rocks must have proceeded more rapidly than at present.

The question then will assume this form. We know that the tides must have been greater in Cambrian or Laurentian days than they are at present; so that they were available as a means of [154] assisting other agents in the stupendous operations of strata manufacture which were then conducted. This certainly helps us to understand how these tremendous beds of strata, a dozen miles or more in solid thickness, were deposited. It seems imperative that for the accomplishment of a task so mighty, some agents more potent than those with which we are familiar should be required. The doctrine of tidal evolution has shown us what those agents were. It only leaves us uninformed as to the degree in which their mighty capabilities were drawn upon.

It is the property of science as it grows to find its branches more and more interwoven, and this seems especially true of the two greatest of all natural sciences—geology and astronomy. With the beginnings of our earth as a globe in the shape in which we find it both these sciences are directly concerned. I have here touched upon another branch in which they illustrate and confirm each other.

As the theory of tidal evolution has shed such a flood of light into the previously dark history of our earth-moon system, it becomes of interest to [155] see whether the tidal phenomena may not have a wider scope; whether they may not, for instance, have determined the formation of the planets by birth from the sun, just as the moon seems to have originated by birth from the earth. Our first presumption, that the cases are analogous, is not however justified when the facts are carefully inquired into. A principle which I have not hitherto discussed here assumes prominence, and therefore we shall devote our attention to it for a few minutes.

Let us understand what we mean by the solar system. There is first the sun at the centre, which preponderates over all the other bodies so enormously, as shown in Fig. 4, in which the earth and the sun are placed side by side for comparison. There is then the retinue of planets, among the smaller of which our earth takes its place, a view of the comparative sizes of the planets being shown in Fig. 5.

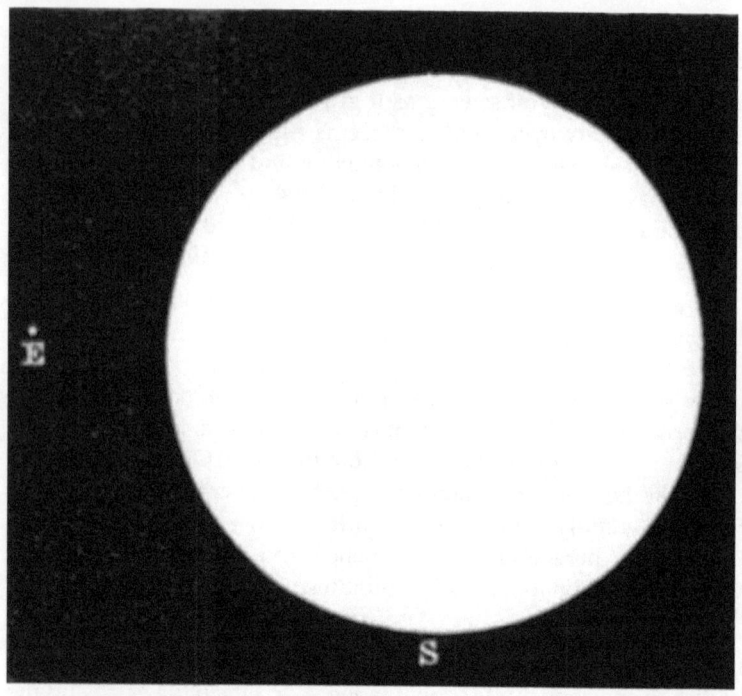

Fig. 4.—Comparative sizes of Earth and Sun.

Not to embarrass ourselves with the perplexities of a problem so complicated as our solar system is in its entirety, we shall for the sake of clear reasoning assume an ideal system, consisting of a sun and [156] a large planet—in fact, such as our own system would be if we could withdraw from it all other bodies, leaving the sun and Jupiter only remaining. We shall suppose, of course, that the sun is much larger than the planet, in fact, it will be convenient to keep in mind the relative masses of the sun and Jupiter, the weight of the planet being less than one-thousandth part of the sun. We know, of course, that both of those bodies are rotating upon [157] their axes, and the one is revolving around the other; and for simplicity we may further suppose that the axes of rotation are perpendicular to the plane of revolution. In bodies so constituted tides [158] will be manifested. Jupiter will raise tides in the sun, the sun will raise tides in Jupiter. If the rotation of each body be performed in a less period than that of the revolution (the case which alone concerns us), then the tides will immediately operate in their habitual manner as a brake for the checking of rotation. The tides raised by the sun on Jupiter will tend therefore to lengthen Jupiter's day; the tides raised on the sun by Jupiter will tend to augment the sun's period of rotation. Both Jupiter and the sun will therefore lose some moment of momentum. We cannot, however, repeat too often the dynamical truth that the total moment of momentum must remain constant, therefore what is lost by the rotation must be made up in the revolution; the orbit of Jupiter around the sun must accordingly be swelling. So far the reasoning appears similar to that which led to such startling consequences in regard to the moon.

Fig. 5.—Comparative sizes of Planets.

But now for the fundamental difference between the two cases. The moon, it will be remembered, always revolves with the same face towards the earth. The tides have ceased to operate there, and consequently the moon is not able to contribute [159] any moment of momentum, to be applied to the enlargement of its distance from the earth; all the moment of momentum necessary for this purpose is of course drawn from the single supply in the rotation of the earth on its axis. But in the case of the system consisting of the sun and Jupiter the circumstances are quite different—Jupiter does not always bend the same face to the sun; so far, indeed, is this from being true, that Jupiter is eminently remarkable for the rapidity of his rotation, and for the incessantly varying aspect in which he would be seen from the sun. Jupiter has therefore a store of available moment of momentum, as has also of course the sun. Thus in the sun and planet system we have in the rotations two available stores of moment of momentum on which the tides can make draughts for application to the enlargement of the revolution. The proportions in which these two available sources can be drawn upon for contributions is not left arbitrary. The laws of dynamics provide the shares in which each of the bodies is to contribute for the joint purpose of driving them further apart.

Let us see if we cannot form an estimate by [160] elementary considerations as to the division of the labour. The tides raised on Jupiter by the sun will be practically proportional to the sun's mass and to the radius of Jupiter. Owing to the enormous size of the sun, the efficiency of these tides and the moment of the friction-brake they produce will be far greater on the planet than will the converse operation of the planet be on the sun. Hence it follows that the efficiency of the tides in depriving Jupiter of moment of momentum will be greatly superior to the efficiency of the tides in depriving the sun of moment of momentum. Without following the matter into any close numerical calculation, we may assert that for every one part the sun contributes to the common object, Jupiter will contribute at least a thousand parts; and this inequality appears all the more striking, not to say unjust, when it is remembered that the sun is more abundantly provided with moment of momentum than is Jupiter—the sun has, in fact, about twenty thousand times as much.

The case may be illustrated by supposing that a rich man and a poor man combine together to achieve some common purpose to which both are to [161] contribute. The ethical notion that Dives shall contribute largely, according to his large means, and Lazarus according to his slender means, is quite antagonistic to the scale which dynamics has imposed. Dynamics declares that the rich man need only give a penny to every pound that has to be extorted from the poor man. Now this is precisely the case with regard to the sun and Jupiter, and it involves a somewhat curious consequence. As long as Jupiter possesses available moment of momentum, we may be certain that no large contribution of moment of momentum has been obtained from the sun. For, returning to our illustration, if we find that Lazarus still has something left in his pocket, we are of course assured that Dives cannot have expended much, because, as Lazarus had but little to begin with, and as Dives only puts in a penny for every pound that Lazarus spends, it is obvious that no large amount can have been devoted to the common object. Hence it follows that whatever transfer of moment of momentum has taken place in the sun-Jupiter system has been almost entirely obtained at the expense of Jupiter. Now in the solar system at present, [162] the orbital moment of momentum of Jupiter is nearly fifty thousand times as great as his present store of rotational moment of momentum. If, therefore, the departure of Jupiter from the sun had been the consequence of tidal evolution, it would follow that Jupiter must once have contained many thousands of times the moment of momentum that he has at present. This seems utterly incredible, for even were Jupiter dilated into an enormously large mass of vaporous matter, spinning round with the utmost conceivable speed, it is impossible that he should ever have possessed enough moment of momentum. We are therefore forced to the conclusion that the tides alone do not provide sufficient explanation for the retreat of Jupiter from the sun.

There is rather a subtle point in the considerations now brought forward, on which it will be necessary for us to ponder. In the illustration of Dives and Lazarus, the contributions of Lazarus of course ceased when his pockets were exhausted, but those of Dives will continue, and in the lapse of time may attain any amount within the utmost limits of Dives' resources. The essential point to [163] notice is, that so long as Lazarus retains anything in *his* pocket, we know for certain that Dives has not given much; if Lazarus, however, has his pocket absolutely empty, and if we do not know how long they may have been in that condition, we have no means of knowing how large a portion of wealth Dives may not have actually expended. The turning-point of the theory thus involves the fact that Jupiter still retains available moment of momentum in his rotation; and this was our sole method of proving that the sun, which in this case was Dives, had never given much. But our argument must have taken an entirely different line had it so happened that Jupiter constantly turned the same face to the sun, and that therefore his pockets were entirely empty in so far as available moment of momentum is concerned. It would be apparently impossible for us to say to what extent the resources of the sun may not have been drawn upon; we can, however, calculate whether in any case the sun could possibly have supplied enough moment of momentum to account for the recession of Jupiter. Speaking in round numbers, the revolutional moment of [164] momentum of Jupiter is about thirty times as great as the rotational moment of momentum at present possessed by the sun. I do not know that there is anything impossible in the supposition that the sun might, by an augmented volume and an augmented velocity of rotation, contain many times the moment of momentum that it has at this moment. It therefore follows that if it had happened that Jupiter constantly bent the same face to the sun, there would apparently be nothing impossible in the fact that Jupiter had been born of the sun, just as the moon was born of the earth. These same considerations should also lead us to observe with still more special attention the development of the earth-moon system. Let us restate the matter of the earth and moon in the light which the argument with respect to Jupiter has given us. At present the rotational moment of momentum of the earth is about a fifth part of the revolutional moment of momentum of the moon. Owing to the fact that the moon keeps the same face to us, she has now no available moment of momentum, and all the moment of momentum required to account for her retreat has [165] of late come from the rotation of the earth; but suppose that the moon still had some liquid on its surface which could be agitated by tides, suppose further that it did not always bend the same face towards us, that it therefore had some available moment of momentum due to its rotation on which the tides could operate, then see how the argument would have been altered. The gradual increase of the moon's distance could be provided for by a transfer of moment of momentum from two sources, due of course to the rotational velocities of the two bodies. Here again the moon and the earth will contribute according to that dynamical but very iniquitous principle which regulated the appropriations from the purses of Dives and Lazarus. The moon must give not according to her abundance, but in the inverse ratio thereof—because she has little she must give largely. Nor shall we make an erroneous estimate if we say that nine-tenths of the whole moment of momentum necessary for the enlargement of the orbit would have been exacted from the moon; that means that the moon must once have had about five or six times as much moment of momentum [166] as the earth possesses at this moment. Considering the small size of the moon, this could only have arisen by terrific velocity of rotation, which it is inconceivable that its materials could ever have possessed.

This presents the demonstration of tidal evolution in a fresh light. If the moon now departed to any considerable extent from showing the constant face to the earth, it would seem that its retreat could not have been caused by tides. Some other agent for producing the present configuration would be necessary, just as we found that some other agent than the tides has been necessary in the case of Jupiter.

But I must say a few words as to the attitude of this question with regard to the entire solar system. This system consists of the sun presiding at the centre, and of the planets and their satellites in revolution around their respective primaries, and each also animated by a rotation on its axis. I shall in so far depart from the actual configuration of the system as to transform it into an ideal system, whereof the masses, the dimensions, and the velocities shall all be preserved; but that the [167] several planes of revolution shall be all flattened into one plane, instead of being inclined at small angles as they are at present; nor will it be unreasonable for us at the same time to bring into parallelism all the axes of rotation, and to arrange that their common directions shall be perpendicular to the plane of their common orbits. For the purpose of our present research this ideal system may pass for the real system.

In its original state, whatever that state may have been, a magnificent endowment was conferred upon the system. Perhaps I may, without derogation from the dignity of my subject, speak of the endowment as partly personal and partly entailed. The system had of course different powers with regard to the disposal of the two portions; the personal estate could be squandered. It consisted entirely of what we call energy; and considering how frequently we use the expression conservation of energy, it may seem strange to say now that this portion of the endowment has been found capable of alienation, nay, further, that our system has been squandering it persistently from the first moment until now. Although the doctrine of the conservation [168] of energy is, we have every reason to believe, a fundamental law affecting the whole universe, yet it would be wholly inaccurate to say that any particular system such as our solar system shall invariably preserve precisely the same quantity of energy without alteration. The circumstance that heat is a form of energy indeed negatives this supposition. For our system possesses energy of all the different kinds: there is energy due to the motions both of rotation and of revolution; there is energy due to the fact that the mutually attracting bodies of our system are separated by distances of enormous magnitude; and there is also energy in the form of heat; and the laws of heat permit that this form of energy shall be radiated off into space, and thus disappear entirely, in so far as our system is concerned. On the other hand, there may no doubt be some small amount of energy accruing to our system from the other systems in space, which like ours are radiating forth energy. Any gain from this source, however, is necessarily so very small in comparison to the loss to which we have referred, that it is quite impossible that the one should balance the other. Though it is undoubtedly [169] true that the total quantity of energy in the universe is constant, yet the share of that energy belonging to any particular system such as ours declines steadily from age to age.

I may indeed remark, that the question as to what becomes of all the radiant energy which the millions of suns in the universe are daily discharging offers a problem apparently not easy to solve; but we need not discuss the matter at present, we are only going to trace out the vicissitudes of our own system; and whatever other changes that system may exhibit, the fact is certain that the total quantity of energy it contains is declining.

Of the two endowments of energy and of moment of momentum originally conferred on our system the moment of momentum is the entailed estate. No matter how the bodies may move, no matter how their actions may interfere with one another, no matter how this body is pulled one way and the other body that way, the conservation of moment of momentum is not imperilled, nor, no matter what losses of heat may be experienced by radiation, could the store of moment of momentum be affected. The only conceivable way in which the [170] quantity of moment of momentum in the solar system could be tampered with is by the interference of some external attracting body. We know, however, that the stars are all situated at such enormous distances, that the influences they can exert in the perturbation of the solar system are absolutely insensible; they are beyond the reach of the most delicate astronomical measurements. Hence we see how the endowment of the system with moment of momentum has conferred upon that system a something which is absolutely inalienable, even to the smallest portion.

Before going any further it would be necessary for me to explain more fully than I have hitherto done the true nature of the method of estimating moment of momentum. The moment of momentum consists of two parts: there is first that due to the revolution of the bodies around the sun; there is secondly the rotation of these bodies on their axes. Let us first think simply of a single planet revolving in a circular orbit around the sun. The momentum of that planet at any moment may be regarded as the product of its mass and its velocity; then the moment of momentum of the [171] planet in the case mentioned is found by multiplying the momentum by the radius of the path pursued. In a more general case, where the planet does not revolve in a circle, but pursued an elliptic path, the moment of momentum is to be found by multiplying the planet's velocity and its mass into the perpendicular from the sun on the direction in which the planet is moving.

These rules provide the methods for estimating all the moments of momentum, so far as the revolutions in our system are concerned. For the rotations somewhat more elaborate processes are required. Let us think of a sphere rotating round a fixed axis. Every particle of that sphere will of course describe a circle around the axis, and all these circles will lie in parallel planes. We may for our present purpose regard each atom of the body as a little planet revolving in a circular orbit, and therefore the moment of momentum of the entire sphere will be found by simply adding together the moments of momentum of all the different atoms of which the sphere is composed. To perform this addition the use of an elaborate mathematical method is required. I do not propose [172] to enter into the matter any further, except to say that the total moment of momentum is the product of two factors—one the angular velocity with which the sphere is turning round, while the other involves the sphere's mass and dimensions.

To illustrate the principles of the computation we shall take one or two examples. Suppose that two circles be drawn, one of which is double the diameter of the other. Let two planets be taken of equal mass, and one of these be put to revolve in one circle, and the other to revolve in the other circle, in such a way that the periods of both revolutions shall be equal. It is required to find the moments of momentum in the two cases. In the larger of the two circles it is plain that the planet must be moving twice as rapidly as in the smaller, therefore its momentum is twice as great; and as the radius is also double, it follows that the moment of momentum in the large orbit will be four times that in the small orbit. We thus see that the moment of momentum increases in the proportion of the squares of the radii. If, however, the two planets were revolving about the same sun, one of these orbits being double the other, the [173] periodic times could not be equal, for Kepler's law tells us that the square of the periodic time is proportional to the cube of the mean distance. Suppose, then, that the distance of the first planet is 1, and that of the second planet is 2, the cubes of those numbers are 1 and 8, and therefore the periodic times of the two bodies will be as 1 to the square root of 8. We can thus see that the velocity of the outer body must be less than that of the inner one, for while the length of the path is only double as large, the time taken to describe that path is the square root of eight times as great; in fact, the velocity of the outer body will be only the square root of twice that of the inner one. As, however, its distance from the sun is twice as great, it follows that the moment of momentum of the outer body will be the square root of twice that of the inner body. We may state this result a little more generally as follows —

In comparing the moments of momentum of the several planets which revolve around the sun, that of each planet is proportional to the product of its mass with the square root of its distance from the sun. [174]

Let us now compare two spheres together, the diameter of one sphere being double that of the other, while the times of rotation of the two are identical. And let us now compare together the moments of momentum in these two cases. It can be shown by reasoning, into which I need not now enter, that the moment of momentum of the large sphere will be thirty-two times that of the small one. In general we may state that if a sphere of homogeneous material be rotating about an axis, its moment of momentum is to be expressed by the product of its angular velocity by the fifth power of its radius.

We can now take stock, as it were, of the constituents of moments of momentum in our system. We may omit the satellites for the present, while such unsubstantial bodies as comets and such small bodies as meteors need not concern us. The present investment of the moment of momentum of our system is to be found by multiplying the mass of each planet by the square root of its distance from the sun; these products for all the several planets form the total revolutional moment of momentum. The remainder of the investment is [175] in rotational moment of momentum, the collective amount of which is to be estimated by multiplying the angular velocity of each planet into its density, and the fifth power of its radius if the planet be regarded as homogeneous, or into such other power as may be necessary when the planet is not homogeneous. Indeed, as the denser parts of the planet necessarily lie in its interior, and have therefore neither the velocity nor the radius of the more superficial portions, it seems necessary to admit that the moments of momentum of the planets will be proportional to some lower power of the radius than the fifth. The total moment of momentum of the planets by rotation, when multiplied by a constant factor, and added to the revolutional moment of momentum, will remain absolutely constant.

It may be interesting to note the present disposition of this vast inheritance among the different bodies of our system. The biggest item of all is the moment of momentum of Jupiter, due to its revolution around the sun; in fact, in this single investment nearly sixty per cent. of the total moment of momentum of the solar system is [176] found. The next heaviest item is the moment of momentum of Saturn's revolution, which is twenty-four per cent. Then come the similar contributions of Uranus and Neptune, which are six and eight per cent. respectively. Only one more item is worth mentioning, as far as magnitude is concerned, and that is the nearly two per cent. that the sun contains in virtue of its *rotation*. In fact, all the other moments of momentum are comparatively insignificant in this method of viewing the subject. Jupiter from his rotation has not the fifty thousandth part of his revolutional moment of momentum, while the earth's rotational share is not one ten thousandth part of that of Jupiter, and therefore is without importance in the general aspect of the system. The revolution of the earth contributes about one eight hundredth part of that of Jupiter.

These facts as here stated will suffice for us to make a forecast of the utmost the tides can effect in the future transformation of our system. We have already explained that the general tendency of tidal friction is to augment revolutional moment of momentum at the expense of rotational. The [177] total, however, of the rotational moment of momentum of the system barely reaches two per cent. of the whole amount; this is of course almost entirely contributed by the sun, for all the planets together have not a thousandth part of the sun's rotational moment. The utmost therefore that tidal evolution can effect in the system is to distribute the two per cent. in augmenting the revolutionary moment of momentum. It does not seem that this can produce much appreciable derangement in the configuration of the system. No doubt if it were all applied to one of the smaller planets it would produce very considerable effect. Our earth, for instance, would have to be driven out to a distance many hundreds of times further than it is at present were the sun's disposable moment of momentum ultimately to be transferred to the earth alone. On the other hand, Jupiter could absorb the whole of the sun's share by quite an insignificant enlargement of its present path. It does not seem likely that the distribution that must ultimately take place can much affect the present configuration of the system.

We thus see that the tides do not appear to [178] have exercised anything like the same influence in the affairs of our solar system generally which they have done in that very small part of the solar system which consists of the earth and moon. This is, as I have endeavoured to show in these lectures, the scene of supremely interesting tidal phenomena; but how small it is in comparison with the whole magnitude of our system may be inferred from the following illustration. I represent the whole moment of momentum of our system by £1,000,000,000, the bulk of which is composed of the revolutional moments of momentum of the great planets, and the rotational moment of momentum of the sun. On this scale the rotational share which has fallen to our earth and moon does not even rise to the dignity of a single pound, it can only be represented by the very modest figure of 19s. 5d. This is divided into two parts—the earth by its rotation accounts for 3s. 4d., leaving 16s. 1d. as the equivalent of the revolution of the moon. The other inferior planets have still less to show than the earth. Venus can barely have more than 2s. 6d.; even Mars' two satellites cannot bring his figure up [179] beyond the slender value of 1½d.; while Mercury will be amply represented by the smallest coin known at her Majesty's mint.

The same illustration will bring out the contrast between the Jovian system and our earth system. The rotational share of the former would be totally represented by a sum of nearly £12,000; of this, however, Jupiter's satellites only contribute about £89, notwithstanding that there are four of them. Thus Jupiter's satellites have not one hundredth part of the moment of momentum which the rotation of Jupiter exhibits. How wide is the contrast between this state of things and the earth-moon system, for the earth does not contain in its rotation one-fifth of the moment of momentum that the moon has in its revolution; in fact, the moon has gradually robbed the earth, which originally possessed 19s. 5d., of which the moon has carried off all but 3s. 4d.

And this process is still going on, so that ultimately the earth will be left very poor, though not absolutely penniless, at least if the retention of a halfpenny can be regarded as justifying that assertion. Saturn, revolving as it does with great [180] rapidity, and having a very large mass, possesses about £2700, while Uranus and Neptune taken together would figure for about the same amount.

In conclusion, let us revert again to the two critical conditions of the earth-moon system. As to what happened before the first critical period, the tides tell us nothing, and every other line of reasoning very little; we can to some extent foresee what may happen after the second critical epoch is reached, at a time so remote that I do not venture even to express the number of ciphers which ought to follow the significant digit in the expression for the number of years. I mentioned, however, that at this time the sun tides will produce the effect of applying a still further brake to the rotation of the earth, so that ultimately the month will have become a shorter period than the day. It is therefore interesting for us to trace out the tidal history of a system in which the satellite revolves around the primary in less time than the primary takes to go round on its own axis—such a system, in fact, as Mars would present at this moment were the outer satellite to be abstracted. The effect of the tides on the planet raised by its satellite [181] would then be to accelerate its rotation; for as the planet, so to speak, lags behind the tides, friction would now manifest itself by the continuous endeavour to drag the primary round faster. The gain of speed, however, thus attained would involve the primary in performing more than its original share of the moment of momentum; less moment of momentum would therefore remain to be done by the satellite, and the only way to accomplish this would be for the satellite to come inwards and revolve in a smaller orbit.

We might indeed have inferred this from the considerations of energy alone, for whatever happens in the deformation of the orbit, heat is produced by the friction, and this heat is lost, and the total energy of the system must consequently decline. Now if it be a consequence of the tides that the velocity of the primary is accelerated, the energy corresponding to that velocity is also increased. Hence the primary has more energy than it had before; this energy must have been obtained at the expense of the satellite; the satellite must therefore draw inwards until it has yielded up enough of energy not alone to account for [182] the increased energy of the primary, but also for the absolute loss of energy by which the whole operation is characterized.

It therefore appears that in the excessively remote future the retreat of the moon will not only be checked, but that the moon may actually return to a point to be determined by the changes in the earth's rotation. It is, however, extremely difficult to follow up the study of a case where the problem of three bodies has become even more complicated than usual.

The importance of tidal evolution in our solar system has also to be viewed in connection with the celebrated nebular hypothesis of the origin of the solar system. Of course it would be understood that tidal evolution is in no sense a rival doctrine to that of the nebular theory. The nebular origin of the sun and the planets sculptured out the main features of our system; tidal evolution has merely come into play as a subsidiary agent, by which a detail here or a feature there has been chiselled into perfect form. In the nebular theory it is believed that the planets and the sun have all originated from the cooling and the contraction of a mighty [183] heated mass of vapours. Of late years this theory, in its main outlines at all events, has strengthened its hold on the belief of those who try to interpret nature in the past by what we see in the present. The fact that our system at present contains some heat in other bodies as well as in the sun, and the fact that the laws of heat require continual loss by radiation, demonstrate that our system, if we look back far enough, and if the present laws have acted, must have had in part, at all events, an origin like that which the nebular theory would suppose.

I feel that I have in the progress of these two lectures been only able to give the merest outline of the theory of tidal evolution in its application to the earth-moon system. Indeed I have been obliged, by the nature of the subject, to omit almost entirely any reference to a large body of the parts of the theory. I cannot bring myself to close these lectures without just alluding to this omission, and without giving expression to the fact, that I feel it is impossible for me to have rendered adequate justice to the strength of the argument on which we claim that tidal evolution is the most rational mode of accounting for the present condition in [184] which we find the earth-moon system. Of course it will be understood that we have never contended that the tides offer the only conceivable theory as to the present condition of things. The argument lies in this wise. A certain body of facts are patent to our observation. The tides offer an explanation as to the origin of these facts. The tides are a *vera causa*, and in the absence of other suggested causes, the tidal theory holds the field. But much will depend on the volume and the significance of the group of associated facts of which the doctrine offers a solution. The facts that it has been in my power to discuss within the compass of discourses like the present, only give a very meagre and inadequate notion of the entire phenomena connected with the moon which the tides will explain. We have not unfrequently, for the sake of simplicity, spoken of the moon's orbit as circular, and we have not even alluded to the fact that the plane of that orbit is inclined to the ecliptic. A comprehensive theory of the moon's origin should render an account of the eccentricity of the moon's orbit; it must also involve the obliquity of the ecliptic, the inclination of the moon's orbit, [185] and the direction of the moon's axis. I have been perforce compelled to omit the discussion of these attributes of the earth-moon system, and in doing so I have inflicted what is really an injustice on the tidal theory. For it is the chief claim of the theory of tidal evolution, as expounded by Professor Darwin, that it links together all these various features of the earth-moon system. It affords a connected explanation, not only of the fact that the moon always turns the same face to the earth, but also of the eccentricity of the moon's path around the earth, and the still more difficult points about the inclinations of the various axes and orbits of the planets. It is the consideration of these points that forms the stronghold of the doctrine of tidal evolution. For when we find that a theory depending upon influences that undoubtedly exist, and are in ceaseless action around us, can at the same time bring into connection and offer a common explanation of a number of phenomena which would otherwise have no common bond of union, it is impossible to refuse to believe that such a theory does actually correspond to nature.

The greatest of mathematicians have ever found [186] in astronomy problems which tax, and problems which greatly surpass, the utmost efforts of which they are capable. The usual way in which the powers of the mathematician have been awakened into action is by the effort to remove some glaring discrepancy between an imperfect theory and the facts of observation. The genius of a Laplace or a Lagrange was expended, and worthily expended, in efforts to show how one planet acted on another planet, and produced irregularities in its orbit; the genius of an Adams and a Leverrier was nobly applied to explain the irregularities in the motion of Uranus, and to discover a cause of those irregularities in the unseen Neptune. In all these cases, and in many others which might be mentioned, the mathematician has been stimulated by the laudable anxiety to clear away some blemish from the theory of gravitation throughout the system. The blemish was seen to exist before its removal was suggested. In that application of mathematics with which we have been concerned in these lectures the call for the mathematician has been of quite a different kind. A certain familiar phenomenon on our sea-coasts [187] has invited attention. The tidal ripples murmur a secret, but not for every ear. To interpret that secret fully, the hearer must be a mathematician. Even then the interpretation can only be won after the profoundest efforts of thought and attention, but at last the language has been made intelligible. The labour has been gloriously rewarded, and an interesting chapter of our earth's history has for the first time been written.

In the progress of these lectures I have sought to interest you in those profound investigations which the modern mathematician has made in his efforts to explore the secrets of nature. He has felt that the laws of motion, as we understand them, are bounded by no considerations of space, are limited by no duration of time, and he has commenced to speculate on the logical consequences of those laws when time of indefinite duration is assumed to be at his disposal. From the very nature of the case, observations for confirmation were impossible. Phenomena that required millions of years for their development [188] cannot be submitted to the instruments in our observatories. But this is perhaps one of the special reasons which make such investigations of peculiar interest, and entitle us to speak of the revelations of Time and Tide as a romance of modern science.

THE END.

www.ingramcontent.com/pod-product-compliance
Lightning Source LLC
Chambersburg PA
CBHW031442210526
45464CB00005B/2301